MATH REFRESHER
FOR SCIENTISTS
AND ENGINEERS

MATH REFRESHER FOR SCIENTISTS AND ENGINEERS

Third Edition

JOHN R. FANCHI

WILEY-INTERSCIENCE

A JOHN WILEY & SONS, INC., PUBLICATION

Published by John Wiley & Sons, Inc., Hoboken, New Jersey
Published simultaneously in Canada

For general information on our other products and services or for technical support, please contact our Customer Care Department within the United States at (800) 762-2974, outside the United States at (317) 572-3993 or fax (317) 572-4002.

Wiley also publishes its books in a variety of electronic formats. Some content that appears in print may not be available in electronic formats. For more information about Wiley products, visit our web site at www.wiley.com.

Library of Congress Cataloging-in-Publication Data.

Fanchi, John R.
  Math refresher for scientists and engineers/John R. Fanchi.—3rd ed.
    p. cm.
  Includes bibliographical references and index.
  ISBN-13: 978-0-471-75715-3
  ISBN-10: 0-471-75715-2
    1. Mathematics. I. Title

  QA37.2.F35 2006
  512'.1—dc22   2005056262

Printed in the United States of America

10 9 8 7 6 5 4 3 2 1

To my sons,
Anthony and Christopher Fanchi

CONTENTS

PREFACE

Math Refresher for Scientists and Engineers, Third Edition is intended for people with technical backgrounds who would like to refresh their math skills. This book is unique because it contains in one source an overview of the essential elements of a wide range of mathematical topics that are normally found in separate texts. The first edition began with relatively simple concepts in college algebra and trigonometry and then proceeded to more advanced concepts ranging from calculus to linear algebra (including matrices) and differential equations. Numerical methods were interspersed throughout the presentation. The second edition added chapters that discussed probability and statistics. In this third edition, three new chapters with exercises and solutions have been added. The new material includes chapters on integral equations, the calculus of variations, and tensor analysis. Furthermore, the discussion of integral transforms has been expanded, a section on partial fractions has been added, and several new exercises have been included.

Math Refresher for Scientists and Engineers, Third Edition is designed for the adult learner and is suitable for reference, self-review, adult education, and college review. It is especially useful for the professional who wants to understand the latest technology, the engineer who is preparing to take a professional engineering exam, or students who wish to refresh their math background. The focus of the book is on modern, practical applications and exercises rather than theory.

Chapters are organized to include a review of important principles and methods. Interwoven with the review are examples, exercises, and applications. Examples are intended to clarify concepts. Exercises are designed to make you an active participant in the review process. Keeping in mind that your time is valuable, I have restricted the number of exercises to a number that should complement and supplement the text while minimizing nonessential repetition. Solutions to exercises are separated from the exercises to allow for a self-paced review.

Applications provide an integration of concepts and methods from one or more sections. In many cases, they introduce you to modern topics. Applications may include material that has not yet been covered in the book, but should be familiar if you are using the book as a "refresher course." You may wish to return to the applications after a first reading of the rest of the text.

I developed much of the material in this book as course notes for continuing education courses in Denver and Houston. I would like to thank Kathy Fanchi, Cindee Calton, Chris Fanchi, and Stephanie Potter for their assistance in the development of this material. Any written comments or suggestions for improving the material are welcome.

JOHN R. FANCHI

CHAPTER 1

ALGEBRA

Many practical applications of advanced mathematics assume the practitioner is fluent in the language of algebra. This review of algebra includes a succinct discussion of sets and groups, as well as a presentation of basic operations with both real and complex numbers. Although much of this material may appear to be elementary to the reader, its presentation here establishes a common terminology and notation for later sections.

1.1 ALGEBRAIC AXIOMS

Sets

A set is a collection of objects. The objects are called *elements* or *members* of the set.

Example: The expression "a and b belong to the set A" can be written $a, b \in A$, where a, b are elements or members of set A.

Let *iff* be the abbreviation for "if and only if." Two sets A, B are equal *iff* they have the same members. In other words, sets A, B satisfy the equality $A = B$ *iff* set A has the same members as set B.

Math Refresher for Scientists and Engineers, Third Edition By John R. Fanchi
Copyright © 2006 John Wiley & Sons, Inc.

Example: Let $A = \{2, 3, 4\}$ and $B = \{4, 5, 6\}$; then $3 \in A$ but $3 \notin B$, where $\notin$ denotes "does not belong to" or "is not a member of." We conclude that set A and set B are not equal.

The union of two sets A and B $(A \cup B)$ is the set of all elements that belong to A or to B or to both. The intersection of two sets A and B $(A \cap B)$ is the set of all elements that belong to both A and B.

Example: Let $A = \{2, 3, 4\}$ and $B = \{4, 5, 6\}$; then $A \cup B = \{2, 3, 4, 5, 6\}$ and $A \cap B = \{4\}$.

Union and intersection may be depicted by a Venn diagram (Figure 1.1). All the elements of a set are enclosed in a circle. The union of the sets is the total area bounded by the circles in the Venn diagram. The intersection is the overlapping cross-hatched area.

The null set or empty set is the set with no elements and is denoted by $\varnothing$.

The set of Real numbers R includes rational numbers and irrational numbers. A rational number is a number that may be written as the quotient of two integers. All other real numbers are irrational numbers. A real number classification scheme is shown as follows:

Rational numbers $\{a/b | a, b \in \text{integers}, b \neq 0\}$
    Noninteger rationals (e.g., $\frac{1}{2}, \frac{1}{3}, \ldots$)
    Integer rationals $\{\ldots, -2, -1, 0, 1, 2, \ldots\}$
        Negative integers $\{\ldots, -3, -2, -1\}$
        Zero $\{0\}$
        Natural numbers $\{1, 2, 3, \ldots\}$
Irrational numbers

A function f from a set A to a set B assigns to each element $x \in A$ a unique element $y \in B$. This may be written $f : A \rightarrow B$ and the function f is said to be a

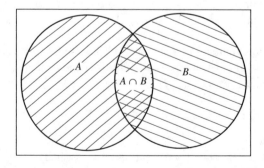

Figure 1.1   Venn diagram.

map of elements from set A to set B. The element y is the value of f at x and is written $y = f(x)$. The set A is the domain of f, and x is an element in the domain of f. The set B is the range of f, and the value of y is in the range of f.

Example: The equation $y = x^2$ is a function that maps the element $x \in R$ to the element $y \in R$. The element y is a function of x, which is often written as $y = f(x)$. By contrast, the equation $y = x^2$ does not uniquely define x as a function of y because x is not single valued; that is, $y = 4$ is associated with both $x = 2$ and $x = -2$. For further discussion of functions, see texts such as Swokowski [1986] or Riddle [1992].

Operations

Square Root For $a \in R$, where a is positive or zero, the notation $\sqrt{a}$ signifies the square root of a. It is the nonnegative real number whose product with itself gives a; thus $\sqrt{a}\sqrt{a} = a$. The notation $\sqrt{a}$ is sometimes called the principal square root [Stewart et al., 1992]. The negative square root of a is $-\sqrt{a}$.

Example: Let $a = 169$; then $\sqrt{a}$ is 13 since $13 \times 13 = 169$. The number $a = 169$ has the negative square root $-\sqrt{169} = -13$.

Absolute Value For all $a \in R$, if a is positive or zero, then $|a| = a$; if a is negative, then $|a| = -a$.

Example: Let $a = \{5, -1, -6, -14.3\}$; then $|a| = \{5, 1, 6, 14.3\}$.

Greater Than Elements can be ordered using the concepts of greater than and less than. For all $a, b \in R$, b is less than a if for some positive real number c we have $a = b + c$. For such a condition, a is said to be greater than b, which may be written as $a > b$. Alternatively, a is greater than b if $a - b > 0$. Similarly, we may say that b is less than a and write it as $b < a$ if $a - b > 0$.

Example: Let $a = 3$ and $b = -2$. Is a greater than b? To satisfy $b + c = a$, we must have $c = 5$. Since c is a positive real number, we know that $a > b$.

Groups

Many of the axioms presented below will seem obvious, particularly when thought of in terms of the set of real numbers. The validity of the axioms becomes much less obvious when applied to different sets, such as sets of matrices.

Axioms   A binary operation ab between a pair of elements $a, b \in G$ exists if $ab \in G$. Let G be a nonempty set with a binary operation. G is a group if the following axioms hold:

[G1]   Associative law: For any $a, b, c \in G$, $(ab)c = a(bc)$.

[G2]   Identity element: There exists an element $e \in G$ such that $ae = ea = a$ for every $a \in G$.

[G3]   Inverse: For each $a \in G$ there exists an element $a^{-1} \in G$ such that $aa^{-1} = a^{-1}a = e$.

Example: Let $a = 3$, $b = 2$, and $c = 5$. We verify the associative law for multiplication by example:

$$(ab)c = a(bc)$$

$$(3 \cdot 2) \cdot 5 = 3 \cdot (2 \cdot 5)$$

$$6 \cdot 5 = 3 \cdot 10$$

$$30 = 30$$

The identity element for multiplication of real numbers is 1, and the inverse of a real number a is $1/a$ as long as $a \neq 0$. The inverse of $a = 0$ is undefined.

Abelian Groups   A group G is an abelian group if $ab = ba$ for every $a, b \in G$. The elements of the group are said to commute. The commutativity of the product of two elements a, b may be written

$$[a, b] \equiv ab - ba = 0$$

Commutativity is always satisfied for the set of real numbers but is often not satisfied for matrices.

Example: Let a and b be the matrices

$$a = \begin{bmatrix} 0 & 1 \\ 1 & 0 \end{bmatrix}, \quad b = \begin{bmatrix} 1 & 0 \\ 0 & -1 \end{bmatrix}$$

Then

$$ab = \begin{bmatrix} 0 & 1 \\ 1 & 0 \end{bmatrix} \begin{bmatrix} 1 & 0 \\ 0 & -1 \end{bmatrix} = \begin{bmatrix} 0 & -1 \\ 1 & 0 \end{bmatrix}$$

$$ba = \begin{bmatrix} 1 & 0 \\ 0 & -1 \end{bmatrix} \begin{bmatrix} 0 & 1 \\ 1 & 0 \end{bmatrix} = \begin{bmatrix} 0 & 1 \\ -1 & 0 \end{bmatrix}$$

Therefore matrices a and b do not commute; that is, $ab \neq ba$.

Hint: To perform the multiplication of the above matrices, recall that

$$\begin{bmatrix} a_{11} & a_{12} \\ a_{21} & a_{22} \end{bmatrix} \begin{bmatrix} b_{11} & b_{12} \\ b_{21} & b_{22} \end{bmatrix} = \begin{bmatrix} a_{11}b_{11} + a_{12}b_{21} & a_{11}b_{12} + a_{12}b_{22} \\ a_{21}b_{11} + a_{22}b_{21} & a_{21}b_{12} + a_{22}b_{22} \end{bmatrix}$$

where $\{a_{ij}\}$ and $\{b_{ij}\}$ are elements of the matrices a and b, respectively. An analogous relation applies when the order of multiplication changes from ab to ba. For more discussion of matrices, see Chapter 4.

Rings

Axioms Let R be a nonempty set with two binary operations:

1. Addition (denoted by $+$), and
2. Multiplication (denoted by juxtaposition).

The nonempty set R is a ring if the following axioms hold:

[R1] Associative law of addition: For any $a, b, c \in R$, $(a + b) + c = a + (b + c)$.

[R2] Zero element: There exists an element $0 \in R$ such that $a + 0 = 0 + a = a$ for every $a \in R$.

[R3] Negative of a: For each $a \in R$ there exists an element $-a \in R$ such that $a + (-a) = (-a) + a = 0$.

[R4] Commutative law of addition: For any $a, b \in R$, $a + b = b + a$.

[R5] Associative law of multiplication: For any $a, b, c \in R$, $(ab)c = a(bc)$.

[R6] Distributive law: For any $a, b, c \in R$, we have

   (i) $a(b + c) = ab + ac$, and

  (ii) $(b + c)a = ba + ca$.

Subtraction is defined in R by $a - b \equiv a + (-b)$. R is a commutative ring if $ab = ba$ for every $a, b \in R$. R is a ring with a unit element if there exists a nonzero element $1 \in R$ such that $a \cdot 1 = 1 \cdot a = a$ for every $a \in R$.

Integral Domain and Field A commutative ring R with a unit element is an integral domain if R has no zero divisors, that is, if $ab = 0$ implies $a = 0$ or $b = 0$. A commutative ring R with a unit element is a field if every nonzero $a \in R$ has a multiplicative inverse; that is, there exists an element $a^{-1} \in R$ such that $aa^{-1} = a^{-1}a = 1$.

1.2 ALGEBRAIC OPERATIONS

Algebraic operations may be presented as a collection of axioms. In all cases assume $a, b, c, d \in R$. The following presents equality axioms:

EQUALITY AXIOMS

Reflexive law	$a = a$
Symmetric law	If $a = b$, then $b = a$
Transitive law	If $a = b$ and $b = c$, then $a = c$
Substitution law	If $a = b$, then a may be substituted for b or b for a in any expression

Ordering relations obey the following axioms:

ORDER AXIOMS

Trichotomy law	Exactly one of the following is true: $a < b$, $a = b$, or $a > b$
Transitive law	If $a < b$ and $b < c$, then $a < c$
Closure for positive numbers	If $a,b > 0$, then $a + b > 0$ and $ab > 0$

Axioms for addition and multiplication operations are summarized as follows:

ADDITION AXIOMS

Closure law for addition	$a + b \in R$
Commutative law for addition	$a + b = b + a$
Associative law for addition	$(a + b) + c = a + (b + c)$
Identity law of addition	$a + 0 = 0 + a = a$
Additive inverse law	$a + (-a) = (-a) + a = 0$

MULTIPLICATION AXIOMS

Closure law for multiplication	$ab \in R$
Commutative law for multiplication	$ab = ba$
Associative law for multiplication	$(ab)c = a(bc)$
Identity law of multiplication	$a \cdot 1 = 1 \cdot a = a$
Multiplication inverse law	$a \cdot (1/a) = (1/a) \cdot a = 1$ for $a \neq 0$
Distributive law	$a \cdot (b + c) = a \cdot b + a \cdot c$

Several algebraic properties follow from the axioms. Some of the most useful are as follows:

MISCELLANEOUS ALGEBRAIC PROPERTIES

1. $a \cdot 0 = 0$.
2. $-(-a) = a$.
3. $-a = -1 \cdot a$.
4. If $a = b$, then $a + c = b + c$.
5. If $a + c = b + c$, then $a = b$.
6. If $a = b$, then $a \cdot c = b \cdot c$.
7. If $a \cdot c = b \cdot c$, then $a = b$ for $c \neq 0$.
8. $a - b = a + (-b)$.
9. $a/b = c/d$ if and only if $a \cdot d = b \cdot c$ for $b, d \neq 0$.
10. $a/b = (a \cdot c)/(b \cdot c)$ for $b, c \neq 0$.
11. $(a/c) + (b/c) = (a + b)/c$ for $c \neq 0$.
12. $(a/b) \cdot (c/d) = (a \cdot c)/(b \cdot d)$ for $b, d \neq 0$.

For a more detailed discussion of algebraic axioms and properties, see such references as Swokowski [1986], Breckenbach et al. [1985], Gustafson and Frisk [1980], and Rich [1973].

1.3 EXPONENTS AND ROOTS

Exponents

Exponents obey the three laws tabulated as follows:

EXPONENTS

Products	$a^m \cdot a^n = a^{m+n}$
Quotient	$\dfrac{a^m}{a^n} = a^{m-n}$ if $m > n$
	$\dfrac{a^m}{a^n} = 1$ if $m = n$
	$\dfrac{a^m}{a^n} = \dfrac{1}{a^{n-m}}$ if $m < n$
Power	$(a^m)^n = a^{mn}$

The number a raised to a negative power is given by $a^{-m} = 1/a^m$. Any nonzero real number raised to the power 0 equals 1; thus $a^0 = 1$ if $a \neq 0$. In the case of the number 0, we have the exponential relationships $0^0 = 0$, $0^x = 0$ for all x.

Example: Scientific notation illustrates the use of exponentiation. In particular, numbers such as 86,400 and 0.00001 can be written in the form 8.64×10^4 and 1×10^{-5}, respectively. Properties of exponents are then used to perform calculations; thus

$$(86,400)(0.00001) = (8.64 \times 10^4)(1 \times 10^{-5})$$
$$= (8.64 \times 1)(10^4 \times 10^{-5})$$
$$= (8.64)(10^{4-5})$$
$$= 8.64 \times 10^{-1} = 0.864$$

Scientific notation is a means of compactly writing very large or very small numbers. It is also useful for making order or magnitude estimates. In this case, numbers are written so that they can be rounded off to approximately 1×10^n, and then exponents are combined to estimate products or quotients. For example, the number 86,400 is approximated as $8.64 \times 10^4 = 0.864 \times 10^5 = 1 \times 10^5$ so that

$$(86,400)(0.00001) \approx (1 \times 10^5)(1 \times 10^{-5}) = 1$$

as expected.

Roots

The solution of the equation

$$b^n = a$$

may be written formally as the nth root of a equals b, or

$$\sqrt[n]{a} = b$$

A real number raised to a fraction p/q may be written as the qth root of a^p, or

$$a^{p/q} = \sqrt[q]{a^p}$$

A summary of relations for roots is as follows:

ROOTS

	$a^{1/x} = \sqrt[x]{a}$
Product	$\sqrt[x]{ab} = \sqrt[x]{a}\sqrt[x]{b} = a^{1/x}b^{1/x}$
Quotient	$\sqrt[x]{\dfrac{a}{b}} = \dfrac{\sqrt[x]{a}}{\sqrt[x]{b}} = \dfrac{a^{1/x}}{b^{1/x}}$
Power	$(a^{1/y})^x = a^{x/y} = \sqrt[y]{a^x}$
	$\sqrt[x]{\sqrt[y]{a}} = (a^{1/y})^{1/x} = a^{1/xy}$

1.4   QUADRATIC EQUATIONS

The solutions of the quadratic equation

$$ax^2 + bx + c = 0$$

are

$$x_{\pm} = \frac{-b \pm \sqrt{b^2 - 4ac}}{2a}$$

where x_+ is the solution when the square root term is added and x_- is the solution when the square root term is subtracted. The quadratic equation is an example of a polynomial equation. Any polynomial equation with a term x^n, where the degree n is the largest positive integer, has n solutions. The quadratic equation has degree $n = 2$ and has two solutions. Cubic equations, which are often encountered in chemical equations of state, have degree $n = 3$ and have three solutions. The solutions are not necessarily real. Polynomial equations are discussed in more detail in Section 1.8.

Example: The solutions of $2x^2 - 6x + 4 = 0$ are

$$x_+ = \frac{6 + \sqrt{(-6)^2 - 4(2)4}}{2(2)} = 2$$

$$x_- = \frac{6 - \sqrt{(-6)^2 - 4(2)4}}{2(2)} = 1$$

These solutions can be used to factor the quadratic equation into a product of terms that have degree $n = 1$; thus

$$2x^2 - 6x + 4 = 2(x - 1)(x - 2) = 0$$

EXERCISE 1.1: Given $ax^2 + bx + c = 0$, find x in terms of a, b, c.

The following exercise may be solved using elementary algebraic operations from the preceding sections.

EXERCISE 1.2: (a) Expand and simplify $(x + y)^2$, $(x - y)^2$, $(x + y)^3$, $(x + y + z)^2$, $(ax + b) \cdot (cx + d)$, and $(x + y)(x - y)$. (b) Factor $3x^3 + 6x^2y + 3xy^2$.

1.5 LOGARITHMS

Definition

The logarithm of the real number $x > 0$ to the base a is written as $\log_a x$ and is defined by the relationship:

$$\text{If } x = a^y \text{ then } y = \log_a x$$

Logarithms are the inverse operation of exponentiation and obey the following three laws:

LOGARITHMS

Product	$\log_a(xy) = \log_a x + \log_a y$
Quotient	$\log_a\left(\dfrac{x}{y}\right) = \log_a x - \log_a y$
Power	$\log_a(x^n) = n\log_a x$
	$\log_a(1) = 0$
	$\log_x(x) = 1$

Logarithms to the base $e \approx 2.71828$ are called *natural logarithms* and are written as $\ln x \equiv \log_e x$. The equation for changing the base of the logarithm from base a to base b is $\log_b x = \log_a x / \log_a b$.

EXERCISE 1.3: Let $y = \log_a x$. Change the base of the logarithm from base a to base b, then set $a = 10$ and $b = 2.71828 \approx e$ and simplify.

Application: Fractals. Benoit Mandelbrot [1967] introduced the concept of fractional dimension, or fractal, to describe the complexities of geographical curves. One of the motivating factors behind this work was an attempt to determine the lengths of coastlines. This problem is discussed here as an introduction to fractals.

We can express the length of a coastline L_c by writing

$$L_c = N\varepsilon \tag{1.5.1}$$

where N is the number of measurement intervals with fixed length ε. For example, ε could be the length of a meter stick. Geographers have long been aware that the length of a coastline depends on its regularity and the unit of length ε. They found that L_c for an irregular coastline, such as the coast of Australia (Figure 1.2), increases as the scale of measurement ε gets shorter. This behavior is caused by the ability of the smaller measurement interval to more accurately include the lengths of irregularities in the measurement of L_c.

Figure 1.2   Irregular boundary.

Richardson [see Mandelbrot, 1983] suggested that N has a power law dependence on ε such that

$$N \propto \varepsilon^{-D} \tag{1.5.2}$$

or

$$N = F\varepsilon^{-D} \tag{1.5.3}$$

where F is a proportionality constant and D is an empirical parameter. Substituting Eq. (1.5.3) into Eq. (1.5.1) gives

$$L(\varepsilon) = F\varepsilon^{(1-D)} \tag{1.5.4}$$

Taking the log of Eq. (1.5.4) yields

$$\log L(\varepsilon) = (1 - D)\log \varepsilon + \log F \tag{1.5.5}$$

If we define $x = \log(\varepsilon)$ and $y = \log L(\varepsilon)$, then Eq. (1.5.5) has the form of the equation of a straight line, $y = mx + b$, where m is the slope and b is the intercept of the straight line with the y axis (see Chapter 3). A plot of $\log(L)$ versus $\log(\varepsilon)$ should yield a straight line with slope $1 - D$.

Richardson applied Eq. (1.5.4) to several geographic examples and found that the power law did not hold [Mandelbrot, 1983, p. 33]. Values of D are given below. Mandelbrot [1967] reinterpreted Richardson's empirical parameter D as a fractional dimension, or fractal. He went on to say that Eq. (1.5.4) applies because a coastline is self-similar; that is, its shape does not depend on the size of the measurement interval. For a more detailed discussion of self-similarity and fractals, see a text such as Devaney [1992].

Example	Fractional Dimension D
West coast of Britain	1.25
German land frontier, A.D. 1900	1.15
Australian coast	1.13
South African coast	1.02
Straight line	1.00

1.6 FACTORIALS

The quantity $n!$ is pronounced "n factorial" and is defined as the product

$$n! = n \cdot (n-1) \cdot (n-2) \cdots (1)$$

By definition, 0! and 1! are equal to 1; thus

$$0! = 1! = 1$$

Example: $4! = 4 \cdot 3 \cdot 2 \cdot 1 = 24$.

Application: Permutations and Combinations: Factorials are used to calculate permutations and combinations. A permutation is an arrangement of distinct objects in a particular order. The number of permutations P_k^n of n distinct objects taken k at a time is

$$P_k^n = \frac{n!}{(n-k)!}$$

If we are not concerned about the order of arrangement of the distinct objects, then we can calculate the number of combinations C_k^n of n distinct objects taken k at a time as

$$C_k^n = \frac{n!}{k!(n-k)!}$$

The number of combinations C_k^n is often written as the binomial coefficient $\binom{n}{k}$. The relationship between permutation P_k^n and combination C_k^n is

$$C_k^n = \frac{P_k^n}{k!}$$

A more detailed discussion of permutations and combinations is presented in several sources, such as Kreyszig [1999], Larsen and Marx [1985], and Mendenhall and Beaver [1991].

EXERCISE 1.4: What are the odds of winning a lotto? To win a lotto, you must select 6 numbers without replacement from a set of 42. Order is not important. Note that the combination of n different things taken k at a time without replacement is

$$\binom{n}{k} = \frac{n!}{k!(n-k)!}$$

Stirling's Approximation

We can approximate the factorial of a large number with the expression

$$n! \approx e^{-n}(n^n)\sqrt{2\pi n} \quad \text{for large } n(n \gg 1)$$

Alternatively, taking the natural logarithm of Stirling's approximation gives

$$\ln n! \approx n \ln n - n + \tfrac{1}{2}\ln(2\pi n)$$

where $\ln x = \log_e x$. The series representation of Stirling's approximation is

$$n! = \sqrt{2\pi n}\, n^n e^{-n}\left[1 + \frac{1}{12n} + \text{smaller terms}\right]$$

EXERCISE 1.5: Calculate $42!/36!$ using Stirling's approximation for both $42!$ and $36!$.

1.7   COMPLEX NUMBERS

If we solve the simple quadratic equation

$$x^2 + 1 = 0$$

we find

$$x^2 = -1$$

which has the solution

$$x = \pm\sqrt{-1} \equiv \pm i$$

This solution requires an extension of the number system to include imaginary numbers, that is, numbers that contain the square root of a negative number as a factor. Another common notation, especially in electrical engineering, for the $\sqrt{-1}$ is $j = \sqrt{-1}$. For consistency, we use $i = \sqrt{-1}$ in this book. It follows from the solution for x that

$$i^2 = -1$$

In general, a complex number is an ordered pair of real numbers (x, y) such that

$$z = x + iy$$

Properties

The real part of a complex number $z = x + iy$ is x, and the imaginary part is y. The real and imaginary parts of a complex number may be written using the notation

$$\text{Re}(z) = x$$
$$\text{Im}(z) = y$$

Complex numbers are often graphed in a two-dimensional plane called the *complex plane* or *Argand diagram* (Figure 1.3).

Two complex numbers

$$z_1 = x_1 + iy_1$$
$$z_2 = x_2 + iy_2$$

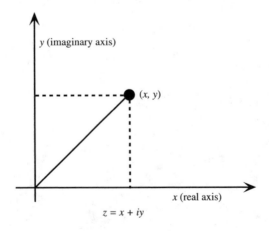

Figure 1.3 Argand diagram (complex plane).

satisfy the equality

$$z_1 = z_2$$

if and only if

$$x_1 = x_2, \quad y_1 = y_2$$

The sum and difference of two complex numbers are

$$z_1 + z_2 = (x_1 + x_2) + i(y_1 + y_2)$$

and

$$z_1 - z_2 = (x_1 - x_2) + i(y_1 - y_2)$$

The product of two complex numbers is

$$z_1 z_2 = (x_1 + iy_1)(x_2 + iy_2)$$
$$= x_1 x_2 + iy_1 x_2 + iy_2 x_1 + i^2 y_1 y_2$$
$$= x_1 x_2 + i(y_1 x_2 + y_2 x_1) - y_1 y_2$$

The conjugate of a complex number z is denoted by z^* and is found by changing i to $-i$ wherever it occurs; thus

$$z_1^* = x_1 - iy_1$$

The complex conjugate z^* is the mirror image of z when the complex numbers z and z^* are plotted in the complex plane illustrated in Figure 1.3. The real x axis acts as the mirror for reflecting one point into another.

Example

$$z_1 z_1^* = (x_1 + iy_1)(x_1 - iy_1) = x_1^2 + y_1^2$$

Division is the inverse of multiplication. The ratio z_1/z_2 of two complex numbers is simplified by calculating $(z_1 z_2^*)/(z_2 z_2^*)$ so that the denominator is real. This operation is illustrated in Exercise 1.6.

EXERCISE 1.6: Factor z_1/z_2 by calculating

$$\frac{z_1 \, z_2^*}{z_2 \, z_2^*} = \frac{z_1 z_2^*}{z_2 z_2^*}$$

Complex numbers satisfy the following laws:

COMPLEX NUMBERS

Commutative law	$z_1 + z_2 = z_2 + z_1$
	$z_1 z_2 = z_2 z_1$
Associative law	$(z_1 + z_2) + z_3 = z_1 + (z_2 + z_3)$
	$(z_1 z_2) z_3 = z_1 (z_2 z_3)$
Distributive law	$z_1 (z_2 + z_3) = z_1 z_2 + z_1 z_3$

The real and imaginary parts of a complex number z may be calculated from z and z^* as

$$\text{Re}(z) = x = \tfrac{1}{2}(z + z^*)$$

and

$$\text{Im}(z) = y = \frac{1}{2i}(z - z^*)$$

respectively. Conjugation of complex numbers satisfies the following frequently encountered relations:

Sum	$(z_1 + z_2)^* = z_1^* + z_2^*$
Difference	$(z_1 - z_2)^* = z_1^* - z_2^*$
Product	$(z_1 z_2)^* = z_1^* z_2^*$
Quotient	$\left(\dfrac{z_1}{z_2}\right)^* = \dfrac{z_1^*}{z_2^*}$

EXERCISE 1.7: Suppose $z = x + iy$ and $z^* = x - iy$. Find x, y in terms of z, z^*.

Application: Particle Lifetimes. The lifetime of some elementary particles may be inferred from measurements of energy distributions. The energy representation may be written as

$$g(\varepsilon) = \frac{i}{\varepsilon + (i\lambda/2)}$$

where ε is particle energy and λ is particle lifetime. The marginal probability density of observing a particle with energy ε and lifetime λ is proportional to $g^*(\varepsilon)g(\varepsilon)$, where

$$g^*(\varepsilon)g(\varepsilon) = \frac{1}{\varepsilon^2 + (\lambda^2/4)}$$

EXERCISE 1.8: Given

$$g = \frac{i}{\varepsilon + (i\lambda/2)}$$

evaluate g^*g.

1.8 POLYNOMIALS AND PARTIAL FRACTIONS

Monomials and Polynomials

A term of the form cx^n, where x is a variable, c is a constant, and n is a positive integer or zero, is called a *monomial function* of x. A function that is the sum of a finite number of monomial terms is called a *polynomial* in x and has the form

$$P(x) = a_n x^n + a_{n-1} x^{n-1} + \cdots + a_1 x + a_0, \quad a_n \neq 0 \qquad (1.8.1)$$

The power n is called the *degree of the polynomial*.

Algebraic and Transcendental Functions   An algebraic function y of x is a function of the form

$$P_0(x)y^n + P_1(x)y^{n-1} + \cdots + P_{n-1}(x)y + P_n(x) = 0 \qquad (1.8.2)$$

where $\{P_i(x): i = 0, 1, \ldots, n\}$ are polynomials in x and n is a positive integer. A function that is not an algebraic function is called a *transcendental* function.

Example: The function $y = \sqrt{x}$ with $x > 0$ is an algebraic function whose elements (x,y) satisfy the equation $y^2 - x = 0$. For this case, $P_0(x) = 1$, $P_1(x) = 0$, and $P_2(x) = -x$ in Eq. (1.8.2) with $n = 2$.

Rational Function   A rational function $R(x)$ is the quotient of two polynomials $N(x)$, $D(x)$ such that

$$R(x) = \frac{N(x)}{D(x)} \qquad (1.8.3)$$

provided that $D(x) \neq 0$. The rational function $R(x)$ is called *proper* if the degree of $N(x)$ is less than the degree of $D(x)$. If the degree of $N(x)$ is greater than the degree of $D(x)$, then $R(x)$ is called *improper*.

Partial Fractions   Suppose the degree of $D(x)$ in Eq. (1.8.3) is less than the degree of $N(x)$ so that $R(x)$ is an improper rational function. In this case, we can divide $N(x)$ by $D(x)$ to obtain a quotient $q(x)$ and a remainder $r(x)$. The resulting ratio is

$$R(x) = \frac{N(x)}{D(x)} = q(x) + \frac{r(x)}{D(x)} \qquad (1.8.4)$$

where the degree of $r(x)$ is less than the degree of $D(x)$. If the degree of $N(x)$ is less than the degree of $D(x)$, then $R(x)$ is a proper rational function and the quotient $q(x) = 0$. The ratio $r(x)/D(x)$ may be decomposed into a sum of simpler terms using the method of partial fractions.

The method of partial fractions is applicable to ratios of polynomials where the degree of $D(x)$ in Eq. (1.8.4) is greater than the degree of $r(x)$. The method for decomposing $r(x)/D(x)$ depends on the form of $D(x)$. Techniques for determining partial fractions are presented in several sources, such as Zwillinger [2003], and Arfken and Weber [2001]. The following examples illustrate different partial fraction techniques.

Example: Consider a polynomial $D(x)$ of degree k. Suppose $D(x)$ may be written as a product of terms that are linear in x so that

$$D(x) = \prod_{i=1}^{k} (a_i x + b_i) = (a_1 x + b_1) \cdot (a_2 x + b_2) \cdots (a_k x + b_k) \qquad (1.8.5)$$

where $\{a_i, b_i\}$ are constants and none of the factors $(a_i x + b_i)$ is duplicated. The ratio $r(x)/D(x)$ can be written as the sum of partial fractions

$$\frac{r(x)}{D(x)} = \sum_{i=1}^{k} \frac{A_i}{a_i x + b_i} \qquad (1.8.6)$$

The constants $\{A_i\}$ are determined by multiplying Eq. (1.8.6) by $D(x)$ and expanding. The resulting equation is a polynomial of degree k. Equating coefficients of powers of x gives a system of k equations for the k unknown constants $\{A_i\}$.

Example: Consider a polynomial $E(x)$ of degree k. Suppose $E(x)$ may be written as a product of terms of the form

$$E(x) = \left[\prod_{i=1}^{k-2} (a_i x + b_i) \right] \times (x^2 + a_k x + b_k) \qquad (1.8.7)$$

where $\{a_i, b_i\}$ are constants. If the factors of the quadratic term $x^2 + a_k x + b_k$ are complex, which occurs when the coefficients $\{a_k, b_k\}$ satisfy the inequality $a_k^2 < 4b_k$, then the partial fraction associated with the quadratic term can be written in the form $(A_k x + B_k)/(x^2 + a_k x + b_k)$. The ratio $r(x)/E(x)$ can be written as the sum of partial fractions

$$\frac{r(x)}{E(x)} = \sum_{i=1}^{k-2} \frac{A_i}{a_i x + b_i} + \frac{A_k x + B_k}{x^2 + a_k x + b_k} \tag{1.8.8}$$

An application of this technique is presented in Exercise 13.4.

EXERCISE 1.9: Write $1/(s^2 + \omega^2)$ in terms of partial fractions.

EXERCISE 1.10: Suppose $r(x) = x + 3$, $D(x) = x^2 + x - 2$. Expand $r(x)/D(x)$ by the method of partial fractions.

CHAPTER 2

GEOMETRY, TRIGONOMETRY, AND HYPERBOLIC FUNCTIONS

A thorough review of mathematics for scientists and engineers would be incomplete without a discussion of the basic concepts of geometry, trigonometry, and hyperbolic functions. Trigonometric and hyperbolic functions are examples of transcendental functions (see Chapter 1, Section 1.8). Like algebra, these topics are essential building blocks for discussing more advanced topics. In addition to definitions, this chapter contains many identities and relations, including definitions of common coordinate systems, Euler's equation, and an introduction to series representations that are useful in subsequent chapters.

2.1 GEOMETRY

Angles

An angle is determined by two rays having the same initial point O as in the sketch.

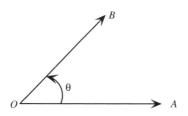

Math Refresher for Scientists and Engineers, Third Edition By John R. Fanchi
Copyright © 2006 John Wiley & Sons, Inc.

The angle θ may be written as angle AOB, and the point O is often referred to as the vertex of the two lines OA and OB. Angles may be measured in degrees or in radians. Each degree may be subdivided into 60 minutes, and each minute may be further subdivided into 60 seconds. If the angle θ equals 360 degrees (denoted $360°$), the line OB will lie on the line OA. There are 2π radians in an angle of $360°$. The value of π is approximately 3.14159.

The following presents terminology for some special angles:

TERMINOLOGY FOR SPECIAL ANGLES

Right angle	$\theta = 90°$
Acute angle	$0° < \theta < 90°$
Obtuse angle	$90° < \theta < 180°$

Two acute angles α, β are complementary if $\alpha + \beta = 90°$. Two positive angles θ, ω are supplementary if $\theta + \omega = 180°$. Some useful angular relations are presented as follows:

ANGULAR RELATIONS

Radians	Degrees
0	$0°$
$\pi/4$	$45°$
$\pi/2$	$90°$
π	$180°$
$3\pi/2$	$270°$
2π	$360°$

Shapes

Right Triangle and Oblique Triangle   The sum of the angles of any triangle in Euclidean geometry must equal $180°$:

$$\alpha + \beta + \gamma = 180°$$

A triangle that does not contain a right angle is called an oblique triangle. A right triangle is a triangle with one angle equal to $90°$, as shown in Figure 2.1. Thus for a right triangle we have the relationship

$$\alpha + \beta = \gamma = 90°$$

between the acute angles α, β. The Pythagorean relation between the lengths of the sides of the right triangle is

$$a^2 + b^2 = c^2$$

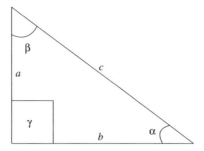

Figure 2.1   Right triangle.

The area of the right triangle is

$$\text{Area} = \frac{ab}{2}$$

Equilateral Triangle   When the lengths of all three sides of a triangle are equal, we have an equilateral triangle. Since the lengths of each side are equal, we also have the angular equality

$$\alpha = \beta = \gamma = 60°$$

and the usual sum

$$\alpha + \beta + \gamma = 180°$$

The area of an equilateral triangle (Figure 2.2) is

$$\text{Area} = \frac{\sqrt{3}}{4} a^2$$

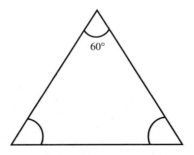

Figure 2.2   Equilateral triangle.

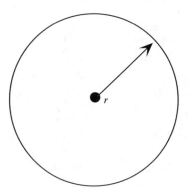

Figure 2.3 Circle.

Circle and Sphere A circle (Figure 2.3) and a sphere with radius r have the following properties:

Circle			Sphere	
Diameter	$2r$		Diameter	$2r$
Circumference	$2\pi r$		Surface Area	$4\pi r^2$
Area	πr^2		Volume	$4\pi r^3/3$

The chord of a circle is a line segment intersecting two points on the circle. The products of the lengths of the segments of any two intersecting chords are related by $ab = cd$, where the lengths are defined by Figure 2.4. Suppose one chord is perpendicular to a diameter so that

$$x(2r - x) = h^2$$

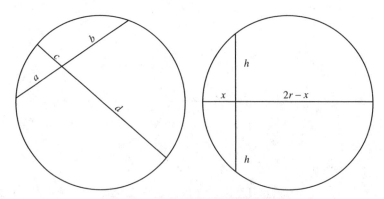

Figure 2.4 Intersecting chords.

If $x << 2r$, we obtain the approximation

$$x \approx \frac{h^2}{2r}$$

which is useful in optics.

Right Circular Cylinder   A right circular cylinder with height h and radius r is shown in Figure 2.5. The right circular cylinder has the following properties:

$$\text{Total Surface Area} = 2\pi rh + 2\pi r^2$$
$$= 2\pi r (h + r)$$
$$\text{Volume} = \pi r^2 h$$

Right Circular Cone   A right circular cone with height h and radius r is shown in Figure 2.6. From the Pythagorean relation for a triangle, we find the slant height s is

$$s = \left(r^2 + h^2\right)^{1/2}$$

The right circular cone has the following properties:

$$\text{Total Surface Area} = \pi rs + \pi r^2$$
$$= \pi r \left[\left(r^2 + h^2\right)^{1/2} + r\right]$$
$$\text{Volume} = \pi r^2 h/3$$

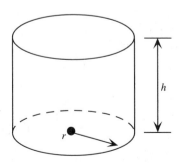

Figure 2.5   Right circular cylinder.

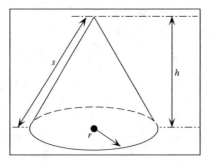

Figure 2.6   Right circular cone.

2.2 TRIGONOMETRY

Trigonometric functions are defined as ratios of the lengths of sides of a right triangle. Using the notation of Figure 2.1, we make the following definitions for an acute angle α:

$$\text{sine } \alpha = \sin \alpha = \frac{a}{c} = \frac{\text{opposite}}{\text{hypotenuse}}$$

$$\text{cosecant } \alpha = \csc \alpha = \frac{c}{a} = \frac{1}{\sin \alpha}$$

$$\text{cosine } \alpha = \cos \alpha = \frac{b}{c} = \frac{\text{adjacent}}{\text{hypotenuse}}$$

$$\text{secant } \alpha = \sec \alpha = \frac{c}{b} = \frac{1}{\cos \alpha}$$

$$\text{tangent } \alpha = \tan \alpha = \frac{a}{b} = \frac{\text{opposite}}{\text{adjacent}}$$

$$\text{cotangent } \alpha = \cot \alpha = \frac{b}{a} = \frac{1}{\tan \alpha}$$

Notice that the sides a, b, c are referred to as the opposite, adjacent, and hypotenuse, respectively. Several useful relations between trigonometric functions can be derived by recalling that the right triangle satisfies the Pythagorean relation $a^2 + b^2 = c^2$. A few of these relations are presented as follows:

TRIGONOMETRIC RELATIONS

Pythagorean	$\sin^2 A + \cos^2 A = 1$
Angle-sum and angle-difference	$\sin (A \pm B) = \sin A \cos B \pm \cos A \sin B$ $\cos (A \pm B) = \cos A \cos B \mp \sin A \sin B$
	$\tan (A \pm B) = \dfrac{\tan A \pm \tan B}{1 \mp \tan A \tan B}$

TRIGONOMETRIC RELATIONS

Double-angle	$\sin 2A = 2 \sin A \cos A$
	$\cos 2A = \cos^2 A - \sin^2 A$
	$= 1 - 2 \sin^2 A$
	$= 2 \cos^2 A - 1$
	$\tan 2A = \dfrac{2 \tan A}{1 - \tan^2 A}$
Product	$\sin A \cos B = \frac{1}{2} \sin(A + B) + \frac{1}{2} \sin(A - B)$
	$\sin A \cos B = \frac{1}{2} \cos(A - B) - \frac{1}{2} \cos(A + B)$
	$\cos A \cos B = \frac{1}{2} \cos(A - B) + \frac{1}{2} \cos(A + B)$
Power	$\sin^2 A = \frac{1}{2}(1 - \cos 2A)$
	$\cos^2 A = \frac{1}{2}(1 + \cos 2A)$
	$\tan^2 A = \dfrac{1 - \cos 2A}{1 + \cos 2A}$
Half-angle	$\sin^2\left(\dfrac{A}{2}\right) = \dfrac{1}{2}(1 - \cos A)$
	$\cos^2\left(\dfrac{A}{2}\right) = \dfrac{1}{2}(1 + \cos A)$
	$\tan^2\left(\dfrac{A}{2}\right) = \dfrac{1 - \cos A}{1 + \cos A}$
Euler	$e^{iA} = \cos A + i \sin A, \; i = \sqrt{-1}$
	$\sin A = \dfrac{1}{2i}\left(e^{iA} - e^{-iA}\right)$
	$\cos A = \dfrac{1}{2}\left(e^{iA} + e^{-iA}\right)$
	$\tan A = -i\left(\dfrac{e^{iA} - e^{-iA}}{e^{iA} + e^{-iA}}\right)$

EXERCISE 2.1: Verify the following relations:

$$\sin^2 \alpha + \cos^2 \alpha = 1$$
$$1 + \tan^2 \alpha = \sec^2 \alpha$$
$$1 + \cot^2 \alpha = \csc^2 \alpha$$
$$\sin \alpha = \tan \alpha \cos \alpha$$
$$\cos \alpha = \cot \alpha \sin \alpha$$
$$\tan \alpha = \frac{\sin \alpha}{\cos \alpha}$$

by rearranging the trigonometric definitions and using the Pythagorean relation.

A positive angle corresponds to a counterclockwise rotation of the hypotenuse from the x axis, and a negative angle corresponds to a clockwise rotation of the hypotenuse from the x axis as illustrated in Figure 2.7.

Based on a geographical analysis and the definitions of trigonometric functions, we find the following relations between trigonometric functions with positive and negative angles:

$$\sin(-\alpha) = -\frac{a}{c} = -\sin \alpha$$
$$\cos(-\alpha) = \frac{b}{c} = \cos \alpha$$
$$\tan(-\alpha) = -\frac{a}{b} = -\tan \alpha$$

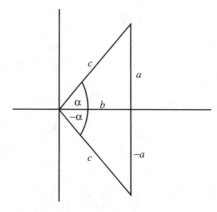

Figure 2.7 Angular rotations.

The numbers a, b in Figure 2.7 are the ordinate (y coordinate) and abscissa (x coordinate), respectively.

Example: We verify that $\tan(-\alpha) = -\tan\alpha$ by using the above relations with the definitions of trigonometric functions. First write

$$\tan(-\alpha) = \frac{\sin(-\alpha)}{\cos(-\alpha)}$$

which, by the above relations, gives the expected result

$$\tan(-\alpha) = \frac{-\sin\alpha}{\cos\alpha} = -\tan\alpha$$

Trigonometric functions are also known as circular functions because of the relationship between the definitions of trigonometric functions and a circle with radius r as sketched in Figure 2.8.

A right triangle is formed by the three sides a, b, r corresponding to the ordinate, abscissa, and radius, respectively. The general definitions of trigonometric functions for an arbitrary angle Θ follow:

$$\sin\Theta = \frac{\text{ordinate}}{\text{radius}} = \frac{a}{r}$$

$$\cos\Theta = \frac{\text{abscissa}}{\text{radius}} = \frac{b}{r}$$

$$\tan\Theta = \frac{\text{ordinate}}{\text{abscissa}} = \frac{a}{b}$$

The radius r is always positive, but the abscissa and ordinate can have positive or negative values. The sign of the trigonometric function depends on the sign of the ordinate and abscissa.

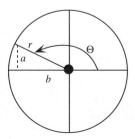

Figure 2.8   General definition of trigonometric functions.

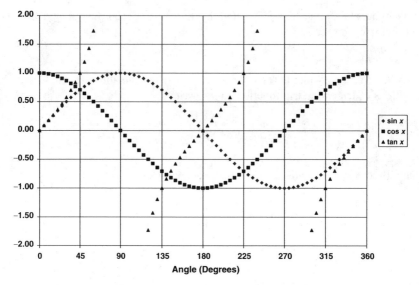

Figure 2.9 Graph of trigonometric functions.

Figure 2.9 shows plots of the trigonometric functions $\sin x$, $\cos x$, and $\tan x$ for the angle x. The value of $\tan x$ is undefined at the angles $x = 90°, 270°$ where the abscissa $b = 0$.

EXERCISE 2.2: *RLC Circuit.* The alternating current $I(t)$ for a resistor–inductor–capacitor circuit with an applied electromotive force is

$$I(t) = -\frac{E_0 S}{S^2 + R^2} \cos(\omega t) + \frac{E_0 R}{S^2 + R^2} \sin(\omega t)$$

Use trigonometric relations to write $I(t)$ in the form

$$I(t) = I_0 \sin(\omega t - \delta)$$

where the maximum current I_0 and phase δ are expressed in terms of E_0, S, and R. A derivation of $I(t)$ is given in Chapter 11, Exercise 11.1.

Law of Cosines

Suppose we want to find the distance C between two points A, B in Figure 2.10. The point A is at $(x_A, y_A) = (b, 0)$ and point B is at $(x_B, y_B) = (a \cos \theta, a \sin \theta)$, where the angle θ is the angle between line OA and line OB. The distance between points A and B is the length C. In general, the length C in Cartesian coordinates is given by

$$C^2 = (x_B - x_A)^2 + (y_B - y_A)^2$$

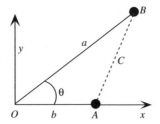

Figure 2.10 Law of cosines.

Substituting the trigonometric values of the Cartesian coordinates gives

$$C^2 = (a \cos \theta - b)^2 + (a \sin \theta)^2$$

Expanding the parenthetic terms

$$C^2 = a^2 \cos^2 \theta - 2ab \cos \theta + b^2 + a^2 \sin^2 \theta$$

and rearranging lets us write

$$C^2 = a^2(\cos^2 \theta + \sin^2 \theta) + b^2 - 2ab \cos \theta$$

Using the identity $\cos^2 \theta + \sin^2 \theta = 1$ gives the law of cosines:

$$C^2 = a^2 + b^2 - 2ab \cos \theta$$

The Pythagorean theorem is obtained when $\theta = 90°$ (Figure 2.11).

Law of Sines

The sides (a, b, c) of the triangle in Figure 2.11 and the sines of the opposite angles (α, β, γ) are related by the law of sines, namely,

$$\frac{a}{\sin \alpha} = \frac{b}{\sin \beta} = \frac{c}{\sin \gamma}$$

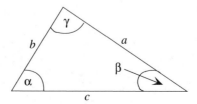

Figure 2.11 Law of sines.

2.3 COMMON COORDINATE SYSTEMS

The location of a point P in three-dimensional space is depicted in Figure 2.12. The coordinates $\{x_1, x_2, x_3\}$ are orthogonal and may be written in terms of Cartesian coordinates $\{x, y, z\}$ as

$$x_1 = x$$
$$x_2 = y$$
$$x_3 = z$$

Cartesian coordinates are also referred to as rectangular coordinates. The range of coordinates in a Cartesian coordinate system is given by

$$-\infty < x < \infty$$
$$-\infty < y < \infty$$
$$-\infty < z < \infty$$

Trigonometric functions can be used to relate the Cartesian coordinate system to two other common coordinate systems: the circular cylindrical coordinate system and the spherical polar coordinate system [e.g., see Arfken and Weber, 2001; Zwillinger, 2003].

Circular Cylindrical Coordinates

The location of point P in circular cylindrical coordinates is depicted in Figure 2.13. The perpendicular distance from the x_3 axis to point P is the cylindrical radius r_c. The angle θ_c is measured in the x_1–x_2 plane from the positive x_1 axis.

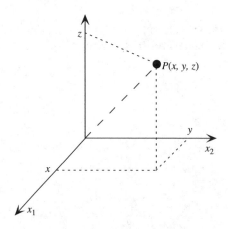

Figure 2.12   Cartesian coordinate system.

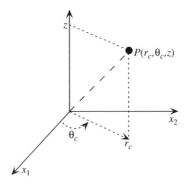

Figure 2.13   Circular cylindrical coordinate system.

The relationship between Cartesian coordinates and circular cylindrical coordinates is

$$x_1 = r_c \cos \theta_c$$
$$x_2 = r_c \sin \theta_c$$
$$x_3 = z$$

The inverse relationship is

$$r_c = \sqrt{x_1^2 + x_2^2}$$
$$\theta_c = \tan^{-1}(x_2/x_1)$$
$$z = x_3$$

The range of coordinates in the circular cylindrical coordinate system is

$$0 \le r_c < \infty$$
$$0 \le \theta_c \le 2\pi$$
$$-\infty < z < \infty$$

Spherical Polar Coordinates

The location of point P in spherical polar coordinates is depicted in Figure 2.14. The distance from the origin to point P is the spherical radius r_s. The azimuth angle ϕ is measured in the x_1–x_2 plane from the positive x_1 axis. The polar angle θ_s is measured from the positive x_3 axis. The relationship between Cartesian coordinates and spherical polar coordinates is

$$x_1 = r_s \sin \theta_s \cos \phi$$
$$x_2 = r_s \sin \theta_s \sin \phi$$
$$x_3 = r_s \cos \theta_s$$

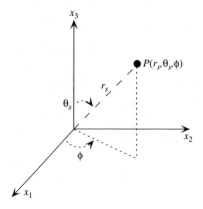

Figure 2.14 Spherical polar coordinate system.

The inverse relationship is

$$r_s = \sqrt{x_1^2 + x_2^2 + x_3^2}$$

$$\theta_s = \cos^{-1}\left(x_3 \middle/ \sqrt{x_1^2 + x_2^2 + x_3^2}\right)$$

$$\phi = \tan^{-1}(x_2/x_1)$$

The range of coordinates in the spherical polar coordinate system is given by

$$0 \le r_s < \infty$$
$$0 \le \theta_s \le \pi$$
$$0 \le \phi \le 2\pi$$

2.4 EULER'S EQUATION AND HYPERBOLIC FUNCTIONS

A fundamental relation between trigonometric and exponential functions is Euler's equation, which has the form

$$e^{i\alpha} = \cos \alpha + i \sin \alpha, \quad i = \sqrt{-1} \tag{2.4.1}$$

Euler's equation can be derived by comparing the series expansions of each function in Eq. (2.4.1). A discussion of series expansions is presented in Section 2.5. Euler's equation for $e^{-i\alpha}$ is

$$e^{-i\alpha} = \cos(-\alpha) + i \sin(-\alpha)$$

or

$$e^{-i\alpha} = \cos \alpha - i \sin \alpha$$

EXERCISE 2.3: Derive expressions for $\cos\alpha$, $\sin\alpha$, and $\tan\alpha$ in terms of exponential functions.

Hyperbolic Functions

Euler's equation may be used to derive several useful relations between trigonometric functions and exponential functions with an imaginary exponent. A new set of functions, called hyperbolic functions, is constructed by using real exponents in relations that are similar to the exponential form of the trigonometric relations. The following presents the definitions and several properties of hyperbolic functions:

HYPERBOLIC FUNCTIONS

Definitions	$\sinh u = \frac{1}{2}(e^u - e^{-u}) = \dfrac{1}{\operatorname{csch} u}$
	$\cosh u = \frac{1}{2}(e^u + e^{-u}) = \dfrac{1}{\operatorname{sech} u}$
	$\tanh u = \dfrac{e^u - e^{-u}}{e^u + e^{-u}} = \dfrac{\sinh u}{\cosh u} = \dfrac{1}{\coth u}$
Product	$\sinh u = \tanh u \cosh u$
	$\cosh u = \coth u \sinh u$
	$\tanh u = \sinh u \operatorname{sech} u$
Angle-sum and angle-difference	$\sinh(u \pm v) = \sinh u \cosh v \pm \cosh u \sinh v$
	$\cosh(u \pm v) = \cosh u \cosh v \pm \sinh u \sinh v$
	$\tanh(u \pm v) = \dfrac{\tanh u \pm \tanh v}{1 \pm \tanh u \tanh v}$
Double-angle	$\sinh 2u = 2 \sinh u \cosh u$
	$\cosh 2u = \cosh^2 u + \sinh^2 u$
	$\tanh 2u = \dfrac{2 \tanh u}{1 + \tanh^2 u}$
Half-angle	$\sinh^2\left(\dfrac{u}{2}\right) = \dfrac{1}{2}(\cosh u - 1)$
	$\cosh^2\left(\dfrac{u}{2}\right) = \dfrac{1}{2}(\cosh u + 1)$
	$\tanh^2\left(\dfrac{u}{2}\right) = \dfrac{\cosh u - 1}{\cosh u + 1}$
Power	$\cosh^2 u - \sinh^2 u = 1$
	$\tanh^2 u + \operatorname{sech}^2 u = 1$

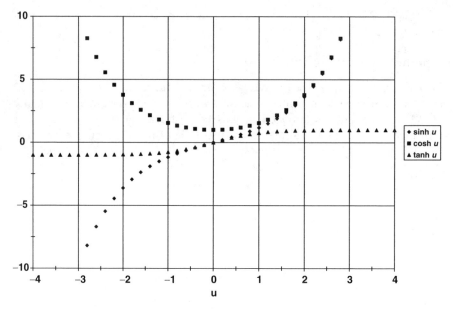

Figure 2.15   Graph of hyperbolic functions.

Relations between hyperbolic functions with a negative argument are

$$\sinh(-u) = -\sinh u$$
$$\cosh(-u) = \cosh u$$
$$\tanh(-u) = -\tanh u$$

Figure 2.15 presents plots of the hyperbolic functions $\sinh u$, $\cosh u$, and $\tanh u$. Notice that $\sinh u$ and $\cosh u$ are approximately equal when $u > 2$ so that $\tanh u \approx 1$ for $u > 2$. By contrast, $\sinh u \approx -\cosh u$ when $u < -2$ so that $\tanh u \approx -1$ when $u < -2$.

EXERCISE 2.4: Use the exponential forms of $\sinh u$ and $\cosh u$ to show that $\sinh(-u) = -\sinh u$ and $\cosh(-u) = \cosh u$.

EXERCISE 2.5: Evaluate $\cosh^2 u - \sinh^2 u$.

Application: Burger's Equation. A nonlinear partial differential equation for the propagation of a shock wave is Burger's equation [Burger, 1948]:

$$\frac{\partial C}{\partial t} = -(a + bC)\frac{\partial C}{\partial x} + D\frac{\partial^2 C}{\partial x^2}$$

If a, b, and D are positive constants, then the solution of Burger's equation is

$$C(x, t) = \frac{1}{2}\left[1 - \tanh\frac{b}{4D}(x - vt)\right]$$

where the velocity v is given by

$$v = a + \frac{b}{2}$$

and the boundary conditions at $x \to \pm\infty$ are $C(-\infty, t) = 1$ and $C(+\infty, t) = 0$. If $b = 0$, then Burger's equation simplifies to the linear convection–dispersion equation. Partial derivatives and partial differential equations are discussed further in Chapters 7 and 12, respectively.

EXERCISE 2.6: Let $z = x - vt$, $v = 1.0$ foot/day, $b = 0.1$ foot/day, and $D = 0.1$ foot^2/day. Plot $C(z)$ in the interval $0 \le x \le 40$ feet at times of 5.00 days, 6.11 days, and 7.22 days. Note that $C(z) = \frac{1}{2}[1 - \tanh(bz/4D)]$ and $0.0 \le C(z) \le 1.0$.

2.5 SERIES REPRESENTATIONS

A polynomial may be written in a power series expansion as

$$P(z) = \sum_{m=0}^{\infty} C_m z^m$$

where $\{C_m, m = 0, \ldots, \infty\}$ are expansion coefficients of the power series. The power series is a polynomial of degree n if it has the form

$$P_n(z) = \sum_{m=0}^{n} C_m z^m, \quad C_n \neq 0$$

Example: The second-degree polynomial $P_2(z) = a + bz + cz^2$ has the expansion coefficients $C_0 = a$, $C_1 = b$, $C_2 = c$, and $C_j = 0$ for all $j > 2$.

Taylor Series and Maclaurin Series

A function $f(z)$ of the complex variable z may be represented by the Taylor series

$$f(z) = \sum_{m=0}^{\infty} \frac{f^{(m)}(a)}{m!}(z - a)^m$$

where

$$f^{(0)}(a) = f(a)$$

The mth derivative of $f(z)$ with respect to z and evaluated at a is

$$f^{(m)}(a) = \frac{d^m f(z)}{dz^m}\bigg|_{z=a}, \quad m \geq 1$$

The center of the Taylor series is at a. If $a = 0$, we obtain the Maclaurin series

$$f_M(z) = \sum_{m=0}^{\infty} \frac{f_M^{(m)}(0)}{m!} z^m$$

For more discussion of derivatives, see Chapter 6.

EXERCISE 2.7: Find the Taylor series and Maclaurin series for $f(z) = e^z$.

Representing Elementary Functions Using Power Series

Examples of power series that are equivalent to several common functions are presented below. If we combine the series for $\cos z$ and $i \sin z$, we obtain Euler's equation (see Section 2.4).

SERIES REPRESENTATION OF ELEMENTARY FUNCTIONS

Geometric series	$\dfrac{1}{1-z} = \displaystyle\sum_{n=0}^{\infty} z^n = 1 + z + z^2 + \cdots, \ |z| < 1$
	The inequality is needed to avoid the singularity (division by 0) at $z = 1$.
Exponential function	$e^z = \displaystyle\sum_{n=0}^{\infty} \dfrac{z^n}{n!} = 1 + z + \dfrac{z^2}{2!} + \cdots$
Trigonometric functions	$\cos z = \displaystyle\sum_{n=0}^{\infty} (-1)^n \dfrac{z^{2n}}{(2n)!} = 1 - \dfrac{z^2}{2!} + \dfrac{z^4}{4!} - + \cdots$
	$\sin z = \displaystyle\sum_{n=0}^{\infty} (-1)^n \dfrac{z^{2n+1}}{(2n+1)!} = z - \dfrac{z^3}{3!} + \dfrac{z^5}{5!} - + \cdots$
Hyperbolic functions	$\cosh z = \displaystyle\sum_{n=0}^{\infty} \dfrac{z^{2n}}{(2n)!} = 1 + \dfrac{z^2}{2!} + \dfrac{z^4}{4!} + \cdots$
	$\sinh z = \displaystyle\sum_{n=0}^{\infty} \dfrac{z^{2n+1}}{(2n+1)!} = z + \dfrac{z^3}{3!} + \dfrac{z^5}{5!} + \cdots$
Logarithm	$\ln(1+z) = \displaystyle\sum_{n=0}^{\infty} \dfrac{z^{n+1}}{n+1}(-1)^n = z - \dfrac{z^2}{2} + \dfrac{z^3}{3} - \cdots$

Given a series representation of a function $f(z)$, it is possible to estimate the asymptotically small values of $f(z)$ as $z \to 0$. This is done by keeping only first-order terms in z, that is, terms with either z^0 or z^1, and neglecting all higher-order terms z^m where $m \geq 2$.

EXERCISE 2.8: Keep only first-order terms in z for small z and estimate asymptotic functional forms for $(1 - z)^{-1}$, e^z, $\cos z$, $\sin z$, $\cosh z$, $\sinh z$, and $\ln (1 + z)$.

Application: Radius of Convergence of Power Series. Consider the power series

$$y(x) = \sum_{n=0}^{\infty} c_n x^n$$

where $\{c_n\}$ are expansion coefficients and x is real. The radius of convergence ρ of the power series $y(x)$ is the limit

$$\rho = \lim_{n \to \infty} \left| \frac{c_n}{c_{n+1}} \right|$$

where c_n and c_{n+1} are consecutive expansion coefficients. Limits are discussed more fully in Chapter 6, Section 6.1. The power series diverges for all values of $x \neq 0$ when $\rho = 0$. If $\rho = \infty$, the power series converges for all values of x. For values of ρ in the interval $0 < \rho < x$, the power series converges when $|x| < \rho$ and diverges when $|x| > \rho$. Additional discussion of power series and convergence criteria can be found in many sources, for example, Lomen and Mark [1988], Edwards and Penney [1989], Kreyszig [1999], and Zwillinger [2003].

Example: All of the expansion coefficients equal 1 for the geometric series. The radius of convergence is therefore $\rho = 1$ for the geometric series, which means the geometric series converges and represents $1/(1 - x)$ when $|x| < 1$.

Example: The expansion coefficients for the exponential series are given by $c_n = 1/n!$ so that the radius of convergence is

$$\rho = \lim_{n \to \infty} \left| \frac{1/n!}{1/(n + 1)!} \right| = \lim_{n \to \infty} |n + 1| = \infty$$

We conclude that the exponential series converges for all values of x.

CHAPTER 3

ANALYTIC GEOMETRY

The review of descriptive geometry in Chapter 2 is extended here to include analytic geometry. The line and conic sections receive special attention because these topics lead naturally to a discussion of linear regression and polar coordinates, respectively. The polar form of complex numbers is then presented.

3.1 LINE

Figure 3.1 illustrates the algebraic representation of a straight line. A straight line is algebraically represented by the equation

$$y = mx + b$$

when m is the slope of the line, and b is the intercept of the line with the y axis. We can find m and b if we know at least two points (x_1, y_1) and (x_2, y_2) on the straight line, as shown in Figure 3.2.

The two unknowns m and b are found by simultaneously solving the two equations

$$y_1 = mx_1 + b \qquad (3.1.1)$$
$$y_2 = mx_2 + b \qquad (3.1.2)$$

Math Refresher for Scientists and Engineers, Third Edition By John R. Fanchi
Copyright © 2006 John Wiley & Sons, Inc.

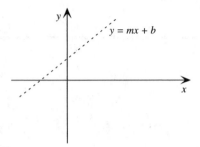

Figure 3.1 Equation of a straight line.

Subtracting Eq. (3.1.1) from Eq. (3.1.2) eliminates b and gives

$$y_2 - y_1 = m(x_2 - x_1) \tag{3.1.3}$$

Solving Eq. (3.1.3) for the slope gives

$$m = \frac{y_2 - y_1}{x_2 - x_1} = \frac{\text{rise}}{\text{run}} \tag{3.1.4}$$

where $y_2 - y_1$ is called the *rise* and $x_2 - x_1$ is called the *run* of the slope m of the straight line. The intercept b is found by rearranging either Eq. (3.1.1) or Eq. (3.1.2). For example, rearranging Eq. (3.1.1) gives

$$b = y_1 - mx_1 \tag{3.1.5}$$

Example: Suppose $(x_1, y_1) = (2, 2), (x_2, y_2) = (-1, -7)$. Using the above questions, we calculate $(b = -4, m = 3)$ and $y = 3x - 4$.

Higher-order polynomial equations may be used to fit three or more data points. A well-defined system will have at least one data point for each unknown coefficient of the polynomial equation.

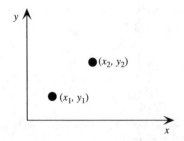

Figure 3.2 Two points of a straight line.

EXERCISE 3.1: Fit the quadratic equation $y = a + bx + cx^2$ to the three points (x_1, y_1), (x_2, y_2), (x_3, y_3).

Application: Least Squares Fit to a Straight Line. We saw above that a straight line could be fit to two points. Suppose we wish to draw the straight line

$$y = mx + b \tag{3.1.6}$$

through N data points. The method of least squares lets us draw a unique straight line through all N data points by minimizing the square of the distance of each point from the straight line. The fit of a straight line to a set of data points using the least squares method is known as *regression analysis* in statistics, and the line in Eq. (3.1.6) is called the *least squares regression line* [Larson et al., 1990].

Any given point (x_i, y_i) is a distance $y_i - mx_i - b$ from the line given by Eq. (3.1.6), as shown in Figure 3.3. The sum of the squares of the distance of each point from the line given by Eq. (3.1.6) is

$$Q = \sum_{i=1}^{N} (y_i - mx_i - b)^2 \tag{3.1.7}$$

The necessary conditions for Q to be a minimum with respect to m and b are that the partial derivatives vanish; that is,

$$\frac{\partial Q}{\partial m} = -2 \sum_{i=1}^{N} x_i (y_i - mx_i - b) = 0 \tag{3.1.8}$$

and

$$\frac{\partial Q}{\partial m} = -2 \sum_{i=1}^{N} (y_i - mx_i - b) = 0 \tag{3.1.9}$$

where we have used the derivative

$$\frac{d(ax^n)}{dx} = nax^{n-1}$$

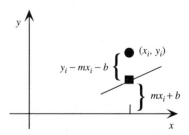

Figure 3.3   Distance of datum from the least squares regression line.

for constants a and n. Derivatives are discussed in Chapter 6 and partial derivatives are discussed in Chapter 7.

Simplifying the summations in Eqs. (3.1.8) and (3.1.9) gives

$$\sum_{i=1}^{N} x_i y_i - m \sum_{i-1}^{N} x_i^2 - b \sum_{i=1}^{N} x_i = 0 \tag{3.1.10}$$

$$\sum_{i=1}^{N} y_i - m \sum_{i=1}^{N} x_i - Nb = 0 \tag{3.1.11}$$

Solving Eq. (3.1.11) for b yields

$$b = -\frac{1}{N}\left(m\sum_{i=1}^{N} x_i - \sum_{i=1}^{N} y_i \right)$$

$$= \frac{1}{N}\left(\sum_{i=1}^{N} y_i - m\sum_{i=1}^{N} x_i \right) \tag{3.1.12}$$

The slope m is found by substituting Eq. (3.1.12) into Eq. (3.1.10):

$$m = \frac{N\left(\sum_{i=1}^{N} x_i y_i\right) - \left(\sum_{i=1}^{N} y_i\right)\left(\sum_{i=1}^{N} x_i\right)}{N\left(\sum_{i=1}^{N} x_i^2\right) - \left(\sum_{i=1}^{N} x_i\right)^2} \tag{3.1.13}$$

Once the slope m is known, Eq. (3.1.12) can be solved for the intercept b.

The least squares analysis may be applied to higher-order polynomials, for example, quadratic or cubic equations. The following exercise shows that the least squares method is applicable to equations that may be transformed into polynomial form.

EXERCISE 3.2: Suppose $w = \alpha u^{\beta}$, where α, β are constants and u, w are variables. Take the logarithm of the equation and fit a least squares regression line to the resulting linear equation.

3.2 CONIC SECTIONS

The most general second-order algebraic equation may be written

$$Ax^2 + Bxy + Cy^2 + Dx + Ey + F = 0 \tag{3.2.1}$$

where the coefficients A, B, C, D, E, and F are constants. Different geometric figures, known as conic sections, correspond to different sets of values of the coefficients.

The circle, ellipse, parabola, and hyperbola are called *conic sections* because each shape can be obtained when a plane intersects one or two right circular cones [Riddle, 1992]. Any conic section can be represented by Eq. (3.2.1) when A, B, and C are not all zero. Conversely, Eq. (3.2.1) is a second-order equation and any second-order equation represents either a conic section or a degenerate case of a conic section. Degenerate cases are summarized as follows [Riddle, 1992] for different values of the product AC of coefficients A and C:

Conic Section	AC	Degenerate Cases
Parabola	$= 0$	One line (two coincident lines)
Ellipse	> 0	Circle
		Point
Hyperbola	< 0	Two intersecting lines

Except for degenerate cases, the graph of Eq. (3.2.1) can be determined by calculating the discriminant

$$\mathcal{D} = B^2 - 4AC$$

The following specifies the relation between a discriminant and its corresponding graph:

Value of Discriminant	Conic Section
$\mathcal{D} = 0$	Parabola
$\mathcal{D} < 0$	Ellipse
$\mathcal{D} > 0$	Hyperbola

The straight line is a degenerate case of Eq. (3.2.1) with $E = 1$, $D = -m$, $F = -b$, and $A = B = C = 0$. Sets of coefficients for several common conic sections are summarized in the following discussions.

Circle

If $A = C = 1$, $F = -r^2$, and $B = D = E = 0$, we obtain

$$x^2 + y^2 = r^2 \quad \text{(constant } r)$$

which is the equation for a circle with radius r.

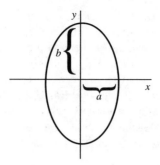

Figure 3.4   Ellipse.

Ellipse

If $A = 1/a^2$, $C = 1/b^2$, $F = -1$, and $B = D = E = 0$, we obtain

$$\frac{x^2}{a^2} + \frac{y^2}{b^2} = 1, \quad b > a$$

which is the equation for an ellipse. Figure 3.4 illustrates an ellipse with $b > a$. The semiminor axis is a and the semimajor axis is b.

Parabola

If $C = 1$, $D = -k$, and $A = B = E = F = 0$, we obtain

$$y^2 = kx$$

which is the equation for a parabola (Figure 3.5).

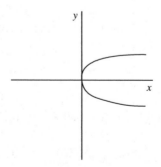

Figure 3.5   Parabola.

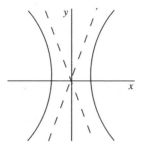

Figure 3.6   Hyperbola.

Hyperbola

If $A = 1/a^2$, $C = -1/b^2$, $F = -1$, and $B = D = E = 0$, we obtain

$$\frac{x^2}{a^2} - \frac{y^2}{b^2} = 1$$

which is the equation for a hyperbola (Figure 3.6). Notice that the only difference between an ellipse and a hyperbola is the sign of C. In general, a parabolic equation has the form $y = x^n$ and $n > 0$, while a hyperbolic equation has the form $y = x^n$ and $n < 0$.

Application: Polar Coordinate Representation of Conic Sections. Another representation of conic sections may be obtained using the polar coordinates sketched in Figure 3.7. The polar coordinate representation is well suited for solving orbit problems in fields such as astrophysics and space science.

Let us first define some terms. In particular, the moving point P is the *parameter*, the fixed point F is the *focus*, and the fixed line D is the *directrix*. The horizontal distance d from the fixed line D to the intersection point I is related to the distance r from the fixed point I by $r = ed$, where e is a constant called the *eccentricity*. The

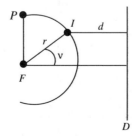

Figure 3.7   Polar coordinates.

length from the focus to the directrix is

$$L_{DF} = d + r \cos v = \frac{r}{e}(1 + e \cos v) \tag{3.2.2}$$

The angle v between the horizontal axis and the line connecting points F and I is called the *true anomaly*.

The parameter P is the length L_{DF} where $v = \pi/2$; that is,

$$r\left(v = \frac{\pi}{2}\right) \equiv P = \frac{eL_{DF}}{1 + e \cos\left(\frac{\pi}{2}\right)} = eL_{DF} \tag{3.2.3}$$

from Eq. (3.2.2). Substituting Eq. (3.2.3) into Eq. (3.2.2) gives

$$r = \frac{eL_{DF}}{1 + e \cos v} = \frac{P}{1 + e \cos v}$$

Parameter P determines the extent of the orbit and e determines the shape. Shapes for different values of e are tabulated as follows:

Eccentricity e	Geometric Shape
0	Circle
$0 < e < 1$	Ellipse
1	Parabola
$e > 1$	Hyperbola

The closest approach of point P to the directrix D occurs at $v = 0$ and is called the *perihelion*. The point P is at its greatest distance from D when $v = \pi$. This distance is called the *aphelion*.

3.3 POLAR FORM OF COMPLEX NUMBERS

In general, a complex number z may be written in terms of the ordered pair of real numbers (x, y) as

$$z = x + iy$$

If we let $x = r \cos \theta$ and $y = r \sin \theta$, where r, θ are polar coordinates, we obtain

$$z = r(\cos \theta + i \sin \theta)$$

The various coordinates are illustrated in Figure 3.8.

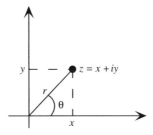

Figure 3.8   Polar form of a complex number.

The length r is the absolute value of modulus of z; thus

$$|z| = r = \sqrt{x^2 + y^2} = \sqrt{zz^*} \geq 0$$

where we have used the definition of the absolute value of z:

$$|z| \equiv \sqrt{zz^*}$$

The angle θ is positive when rotated in the counterclockwise direction and is measured in radians. It is obtained as the argument of z such that

$$\arg z = \theta = \sin^{-1}\frac{y}{r} = \cos^{-1}\frac{x}{r} = \tan^{-1}\frac{y}{x}$$

The value of θ in the interval $-\pi < \theta \leq \pi$ is called the *principal value* of the argument of z.

Euler's Equation

As we saw in Chapter 2, combining the series expansions for $\cos \theta$ and $\sin \theta$ gives Euler's equation:

$$e^{i\theta} \cos \theta + i \sin \theta, 0 \leq \theta \leq 2\pi$$

Multiplying $e^{i\theta}$ by r yields

$$z = re^{i\theta} = r(\cos \theta + i \sin \theta)$$

Comparing the polar coordinate expression of z with the Cartesian coordinate representation $z = x + iy$ gives

$$z = re^{i\theta} = x + iy$$

De Moivre's Equation

If we raise Euler's equation to the nth power we obtain

$$(\cos \theta + i \sin \theta)^n = e^{in\theta}$$

Repeated use of Euler's equation lets us write

$$e^{in\theta} = \cos(n\theta) + i \sin(n\theta) = (\cos \theta + i \sin \theta)^n$$

Multiplying by r^n and using the polar coordinate representation of a number gives De Moivre's equation:

$$z^n = r^n e^{in\theta}$$

EXERCISE 3.3: Express $z = z_1 z_2$ and $z = z_1/z_2$ in polar coordinates.

CHAPTER 4

LINEAR ALGEBRA I

The introduction of polar coordinates in Chapter 3 established the basis for discussing the rotation of coordinate axes in two dimensions. This coordinate rotation requires the solution of two equations for two unknowns. The need to solve the resulting set of equations leads us to a discussion of matrices and determinants.

4.1 ROTATION OF AXES

Figure 4.1 depicts a set of coordinates (x_1, x_2) undergoing a counterclockwise rotation through an angle θ to a new set of coordinates (y_1, y_2). We wish to describe the location of the point P relative to both sets of coordinates.

In the (x_1, x_2) coordinate system, the point P is at

$$x_1 = r \cos(\phi)$$
$$x_2 = r \sin(\phi) \tag{4.1.1}$$

whereas in the (y_1, y_2) coordinate system, P is at

$$y_1 = r \cos(\phi - \theta)$$
$$y_2 = r \sin(\phi - \theta) \tag{4.1.2}$$

Math Refresher for Scientists and Engineers, Third Edition By John R. Fanchi
Copyright © 2006 John Wiley & Sons, Inc.

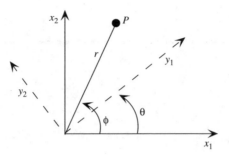

Figure 4.1   Coordinate rotation.

Using the trigonometric identities

$$\sin (A \pm B) = \sin A \cos B \pm \cos A \sin B$$
$$\cos (A \pm B) = \cos A \cos B \mp \sin A \sin B \tag{4.1.3}$$

in Eq. (4.1.2)

$$\begin{aligned}
y_1 &= r(\cos \phi \cos \theta + \sin \phi \sin \theta) \\
&= (r \cos \phi) \cos \theta + (r \sin \phi) \sin \theta \\
y_2 &= r(\sin \phi \cos \theta - \cos \phi \sin \theta) \\
&= (r \sin \phi) \cos \theta - (r \cos \phi) \sin \theta
\end{aligned} \tag{4.1.4}$$

Substituting Eq. (4.1.1) into Eq. (4.1.4) gives a system of linear equations relating the two coordinate systems in terms of the angle of rotation θ; thus

$$\begin{aligned}
y_1 &= x_1 \cos \theta + x_2 \sin \theta \\
y_2 &= x_2 \cos \theta - x_1 \sin \theta
\end{aligned} \tag{4.1.5}$$

The preceding set of equations represents a transformation from the (x_1, x_2) coordinate system to the (y_1, y_2) coordinate system. The set of equations in Eq. (4.1.5) may be written in a more compact form as

$$\begin{bmatrix} y_1 \\ y_2 \end{bmatrix} = \begin{bmatrix} \cos \theta & \sin \theta \\ -\sin \theta & \cos \theta \end{bmatrix} \begin{bmatrix} x_1 \\ x_2 \end{bmatrix} \tag{4.1.6}$$

The equation may be written in an even more compact matrix notation as

$$y = ax$$

where x, y are column vectors and a is a 2×2 matrix that transforms x into y. Matrices appear in numerous real-world applications. The rest of this chapter focuses on their properties and serves as an introduction to linear algebra. More detailed treatments of linear algebra can be found in such sources as Berberian [1992], Nicholson [1986], and Rudin [1991].

4.2 MATRICES

A matrix is an array of numbers with m rows and n columns. The $(i, j)^{\text{th}}$ element is the element found in row i and column j. Matrices may be classified by the relationship between m and n (see below). A matrix is often referred to by its order, for example, the matrix is an $m \times n$ matrix.

Rectangular matrix	$m \neq n$
Square matrix	$m = n$
Order of matrix	$m \times n$

Example: The matrix has $m = 2$ rows, $n = 3$ columns, and order 2×3. The $(i, j)^{\text{th}}$ element is a_{ij}.

$$A = \begin{bmatrix} a_{11} & a_{12} & a_{13} \\ a_{21} & a_{22} & a_{23} \end{bmatrix}$$

A column vector is a matrix of order $m \times 1$, whereas a row vector is a matrix of order $1 \times n$. A number may be thought of as a matrix of order 1×1 and is considered a scalar if its value is invariant (does not change) with respect to a coordinate transformation.

Matrices may be classified further based on the properties of their elements. The transpose of matrix A is formed by interchanging element a_{ij} with element a_{ji}; thus

$$A^T = [a_{ji}] = [a_{ij}]^T = (A)^T$$

where A^T denotes the transpose of matrix A.

Example: The transpose of the 2×3 matrix A from the previous example is the 3×2 matrix:

$$A^T = \begin{bmatrix} a_{11} & a_{21} \\ a_{12} & a_{22} \\ a_{13} & a_{23} \end{bmatrix}$$

Example: A 3×1 column vector is

$$a = \begin{bmatrix} a_1 \\ a_2 \\ a_3 \end{bmatrix}$$

The transpose of a is the 1×3 row vector $a^T = [a_1 \; a_2 \; a_3]$.

The conjugate transpose or adjoint of matrix A is

$$A^+ = [A^*]^T$$

where A^* is the complex conjugate of A or

$$[a_{ij}]^+ = [a_{ij}^*]^T = [a_{ji}^*]$$

A square matrix is symmetric if $A = A^T$; that is, $a_{ij} = a_{ji}$. A square matrix is Hermitian if $A = A^+$. In this case the matrix is said to be *self-adjoint*. An important, physically useful property of Hermitian matrices is that they have real eigenvalues (see Chapter 5).

EXERCISE 4.1: The Pauli matrices are the square matrices

$$\sigma_x = \begin{bmatrix} 0 & 1 \\ 1 & 0 \end{bmatrix}, \quad \sigma_y = \begin{bmatrix} 0 & -i \\ i & 0 \end{bmatrix}, \quad \sigma_z = \begin{bmatrix} 1 & 0 \\ 0 & -1 \end{bmatrix}$$

These matrices are used to calculate the spin angular momentum of particles such as electrons, protons, and neutrons. Show that the Pauli matrices are Hermitian.

The principal diagonal of a square matrix A with elements $\{a_{ij}\}$ is the set of elements a_{ij} for which $i = j$. The trace of an $n \times n$ square matrix A is the sum of the elements along the principal diagonal; thus

$$\text{Trace}(A) \equiv \text{Tr}(A) = \sum_{i=1}^{n} a_{ii}$$

If $a_{ij} = 0$ for $i < j$, then A is a lower triangular matrix. If $a_{ij} = 0$ for $i > j$, then A is an upper triangular matrix. A diagonal matrix is a square matrix with $a_{ij} = 0$ for $i \neq j$ and $a_{ii} \neq 0$. The elements a_{ij} with $i \neq j$ are called *off-diagonal elements*.

Example: Consider the 3×3 square matrix A:

$$A = \begin{bmatrix} a_{11} & a_{12} & a_{13} \\ a_{21} & a_{22} & a_{23} \\ a_{31} & a_{32} & a_{33} \end{bmatrix}$$

The sum of the elements of the principal diagonal of A is the trace of A:

$$\text{Tr}(A) = \sum_{i=1}^{3} a_{ii} = a_{11} + a_{22} + a_{33}$$

If all off-diagonal elements of A are 0, that is, $a_{ij} = 0$ for $i \neq j$, then we have the diagonal matrix

$$\text{diag } A = \begin{bmatrix} a_{11} & 0 & 0 \\ 0 & a_{22} & 0 \\ 0 & 0 & a_{33} \end{bmatrix}$$

Upper and lower triangular matrices U, L may be formed from A and written as

$$U = \begin{bmatrix} a_{11} & a_{12} & a_{13} \\ 0 & a_{22} & a_{23} \\ 0 & 0 & a_{33} \end{bmatrix}, \quad L = \begin{bmatrix} a_{11} & 0 & 0 \\ a_{21} & a_{22} & 0 \\ a_{31} & a_{32} & a_{33} \end{bmatrix}$$

In general, a matrix is an upper triangular matrix when all the entries below the main diagonal are zero. Similarly, the matrix is a lower triangular matrix when all the entries above the main diagonal are zero.

Matrix Operations

Assume matrices A, B, C are matrices with the same order $m \times n$. The sum or difference of two matrices A, B is

$$A \pm B = C$$

which implies

$$a_{ij} \pm b_{ij} = c_{ij} \quad \text{for } i = 1, 2, \ldots, m$$
$$j = 1, 2, \ldots, m$$

The product of a matrix A with a scalar γ is

$$B = \gamma A$$

or, in terms of matrix elements,

$$b_{ij} = \gamma a_{ij}$$

Example**:** Define a matrix A with complex elements such that

$$A = \begin{bmatrix} 0 & i \\ -i & 0 \end{bmatrix}$$

Then $6A$ and $3i\,A$ are given by

$$6A = \begin{bmatrix} 0 & 6i \\ -6i & 0 \end{bmatrix}, \quad 3i\,A = \begin{bmatrix} 0 & -3 \\ 3 & 0 \end{bmatrix}$$

Multiplication   Let A be of order $m \times n$ and B of order $n \times p$. Then the product of two matrices A and B is

$$C = AB$$

or

$$c_{ij} = \sum_{h=1}^{n} a_{ih} b_{hj}$$

where the resulting matrix C is of order $m \times p$.

EXERCISE 4.2: Calculate

$$\begin{bmatrix} 3 & 4 & 2 \\ 2 & 3 & -1 \end{bmatrix} \begin{bmatrix} 1 & -2 & -4 \\ 0 & -1 & 2 \\ 6 & -3 & 9 \end{bmatrix}$$

EXERCISE 4.3: Calculate

$$\begin{bmatrix} 2 & 3 & -1 \end{bmatrix} \begin{bmatrix} -4 \\ 2 \\ 9 \end{bmatrix}$$

Square matrices obey the laws expressed as follows:

Associative	$A(BC) = (AB)C$
Distributive	$(A + B)C = AC + BC$ and
	$C(A + B) = CA + CB$

In general, matrix multiplication is not commutative; that is, $AB \neq BA$. This is an important characteristic of quantum mechanics.

Example: Compare $\sigma_x \sigma_y$ and $\sigma_y \sigma_x$.

$$\sigma_x\sigma_y = \begin{bmatrix} 0 & 1 \\ 1 & 0 \end{bmatrix}\begin{bmatrix} 0 & -i \\ i & 0 \end{bmatrix} = \begin{bmatrix} i & 0 \\ 0 & -i \end{bmatrix}$$

$$\sigma_y\sigma_x = \begin{bmatrix} 0 & -i \\ i & 0 \end{bmatrix}\begin{bmatrix} 0 & 1 \\ 1 & 0 \end{bmatrix} = \begin{bmatrix} -i & 0 \\ 0 & i \end{bmatrix}$$

Notice that $\sigma_x \sigma_y \neq \sigma_y \sigma_x$; that is, σ_x and σ_y do not commute.

The transpose of the product of two matrices is the product of the commuted, transposed matrices:

$$(AB)^T = B^T A^T$$

The adjoint of the product of two matrices is the product of the commuted, adjoint matrices:

$$(AB)^+ = B^+ A^+$$

If all of the nonzero elements of a square, diagonal matrix equal 1, then that matrix is called the *identity matrix*. It satisfies the equality

$$AI = IA = A$$

where A is any square matrix with the same order as the square identity matrix I. The matrix I has the form

$$I = \begin{bmatrix} 1 & 0 & \cdots & 0 \\ 0 & 1 & \cdots & 0 \\ \vdots & & \ddots & \\ 0 & 0 & \cdots & 1 \end{bmatrix}$$

A null matrix (**0**) is a matrix in which all elements equal 0; thus $A0 = 0A = 0$. A matrix A is singular if

$$Ax = 0, x \neq 0$$

where x is a column vector [Lipschutz, 1968]. Notice that $x \neq 0$ if at least one element of x is nonzero. Alternatively, it is possible to show that a matrix A is singular if its determinant is zero. This is pointed out again in Section 4.3 after determinants have been defined.

Suppose A is a square matrix and there exists a square matrix B such that $AB = I$ or $BA = I$. Then A is called a nonsingular matrix and B is the inverse of matrix A; that is, $B = A^{-1}$, where A^{-1} denotes the inverse of matrix A. Inverse matrices are discussed in more detail below.

As with scalars, we can define the power of a matrix by forming the products

$$A^2 \equiv AA, \quad A^3 \equiv AAA, \text{etc.}$$

In addition, a matrix may be used as an exponent by using the series expansion

$$e^A \equiv \sum_{m=0}^{\infty} \frac{A^m}{m!}$$

EXERCISE 4.4: Suppose we are given the Pauli matrices

$$\sigma_1 = \begin{bmatrix} 0 & 1 \\ 1 & 0 \end{bmatrix}, \quad \sigma_2 = \begin{bmatrix} 0 & -i \\ i & 0 \end{bmatrix}, \quad \sigma_3 = \begin{bmatrix} 1 & 0 \\ 0 & -1 \end{bmatrix}$$

(a) Calculate σ_3^2. (b) Calculate the trace of σ_1 and the trace of σ_3. (c) Verify the equality $[\sigma_2 \, \sigma_3]^T = \sigma_3^T \sigma_2^T$.

Direct Product of Matrices

Suppose we have an $m \times m$ matrix A and an $n \times n$ matrix B. We can form an $mn \times mn$ matrix C by defining the direct product

$$C = A \otimes B = \begin{bmatrix} a_{11}B & a_{12}B & \cdots & a_{1m}B \\ a_{21}B & a_{22}B & \cdots & a_{2m}B \\ \vdots & \vdots & & \vdots \\ a_{m1}B & a_{m2}B & \cdots & a_{mm}B \end{bmatrix}$$

To be more specific, let A and B be the 2×2 matrices

$$A = \begin{bmatrix} a_{11} & a_{12} \\ a_{21} & a_{22} \end{bmatrix}, \quad B = \begin{bmatrix} b_{11} & b_{12} \\ b_{21} & b_{22} \end{bmatrix}$$

The direct product matrix C is the 4×4 matrix

$$C = A \otimes B = \begin{bmatrix} a_{11}B & a_{12}B \\ a_{21}B & a_{22}B \end{bmatrix} = \begin{bmatrix} a_{11}b_{11} & a_{11}b_{12} & a_{12}b_{11} & a_{12}b_{12} \\ a_{11}b_{21} & a_{11}b_{22} & a_{12}b_{21} & a_{12}b_{22} \\ a_{21}b_{11} & a_{21}b_{12} & a_{22}b_{11} & a_{22}b_{12} \\ a_{21}b_{21} & a_{21}b_{22} & a_{22}b_{21} & a_{22}b_{22} \end{bmatrix}$$

In general, the direct product is not necessarily commutative; that is, $A \otimes B$ may not equal $B \otimes A$.

Example: Let **0** be a 2×2 matrix with all of its elements equal to 0, and

$$A = \boldsymbol{\sigma}_x = \begin{bmatrix} 0 & 1 \\ 1 & 0 \end{bmatrix}, \; B = \boldsymbol{\sigma}_y = \begin{bmatrix} 0 & -i \\ i & 0 \end{bmatrix}$$

The direct product is

$$\boldsymbol{\sigma}_x \otimes \boldsymbol{\sigma}_y = \begin{bmatrix} \mathbf{0} & \boldsymbol{\sigma}_y \\ \boldsymbol{\sigma}_y & \mathbf{0} \end{bmatrix} = \begin{bmatrix} 0 & 0 & 0 & -i \\ 0 & 0 & i & 0 \\ 0 & -i & 0 & 0 \\ i & 0 & 0 & 0 \end{bmatrix}$$

We show that the direct product does not necessarily commute by calculating

$$\boldsymbol{\sigma}_y \otimes \boldsymbol{\sigma}_x = \begin{bmatrix} \mathbf{0} & -i\boldsymbol{\sigma}_x \\ i\boldsymbol{\sigma}_x & \mathbf{0} \end{bmatrix} \neq \boldsymbol{\sigma}_x \otimes \boldsymbol{\sigma}_y$$

Inverse Matrices

If A is a nonsingular square matrix of order $n \times n$, then there exists a unique matrix A^{-1} such that

$$A^{-1}A = AA^{-1} = I$$

The matrix A^{-1} is said to be the *inverse* of matrix A.

The inverse of the product of two matrices is the product of the commuted, inverse matrices:

$$(AB)^{-1} = B^{-1}A^{-1}$$

The transpose of the inverse matrix is the inverse of the transposed matrix:

$$(A^{-1})^T = (A^T)^{-1}$$

The inverse of a matrix multiplied by a scalar gives

$$(\gamma A)^{-1} = \frac{1}{\gamma}A^{-1}$$

A square matrix A is said to be a unitary matrix if its inverse is self-adjoint; that is,

$$A^+ = A^{-1}$$

Two matrices P, Q are similar if

$$P = A^{-1}QA$$

where A is a nonsingular matrix.

EXERCISE 4.5: Let

$$A = \begin{bmatrix} 2 & 5 \\ 1 & 3 \end{bmatrix}, \ A^{-1} = \begin{bmatrix} 3 & -5 \\ -1 & 2 \end{bmatrix}$$

Verify $AA^{-1} = I$.

EXERCISE 4.6: Let

$$A = \begin{bmatrix} 1 & 0 & 2 \\ 2 & -1 & 3 \\ 4 & 1 & 8 \end{bmatrix}, \ A^{-1} = \begin{bmatrix} -11 & 2 & 2 \\ -4 & 0 & 1 \\ 6 & -1 & -1 \end{bmatrix}$$

Verify $AA^{-1} = I$.

EXERCISE 4.7: Let

$$A = \begin{bmatrix} a & b \\ c & d \end{bmatrix}, \ A^{-1} = \begin{bmatrix} x & y \\ z & w \end{bmatrix}$$

Assume a, b, c, d are known. Find x, y, z, w from $AA^{-1} = I$.

Solving a System of Linear Equations

A system of n linear equations with n unknowns can be written in the form

$$Ax = b \tag{4.2.1}$$

where the column vector b is known and the column vector x is unknown. We assume the matrix A is a known, nonsingular square matrix of order $n \times n$ so that there exists a unique inverse matrix A^{-1}. The formal solution for x is obtained by premultiplying Eq. (4.2.1) by A^{-1} to get

$$x = A^{-1}b \tag{4.2.2}$$

Notice that multiplying a matrix equation by a matrix from the left (premultiplying) is not necessarily equivalent to multiplying the matrix equation by a matrix from the right (postmultiplying) because matrix multiplication does not have to commute, and often does not.

EXERCISE 4.8: Assume b_1, b_2 are known constants. Write the system of equations

$$2x_1 + 5x_2 = b_1$$
$$x_1 + 3x_2 = b_2$$

in matrix form and solve for the unknowns x_1, x_2.

4.3 DETERMINANTS

The determinant of the square matrix A is denoted by $\det(A)$ or $|A|$. $|A|$ is a scalar function of the square matrix A defined such that

$$|A|\,|B| = |AB|$$

The determinant of the transposed of matrix A is the same as the determinant of A; thus

$$|A| = |A^T|$$

An explicit representation of the determinant of a 2×2 matrix is

$$\det\begin{bmatrix} a_{11} & a_{12} \\ a_{21} & a_{22} \end{bmatrix} = a_{11}a_{22} - a_{21}a_{12}$$

Similarly, the determinant of a 3×3 matrix is

$$\det\begin{bmatrix} a_{11} & a_{12} & a_{13} \\ a_{21} & a_{22} & a_{23} \\ a_{31} & a_{32} & a_{33} \end{bmatrix}$$
$$= [a_{11}a_{22}a_{33} + a_{12}a_{23}a_{31} + a_{13}a_{21}a_{32}$$
$$- a_{31}a_{22}a_{13} - a_{32}a_{23}a_{11} - a_{33}a_{21}a_{12}]$$

The determinant of a 3×3 matrix is found operationally by appending the first two columns on the right-hand side of the matrix and subtracting the products of the diagonal terms (multiply terms from lower left to upper right)

$$
\begin{array}{ccccc}
a_{11} & a_{12} & a_{13} & a_{11} & a_{12} \\
a_{21} & a_{22} & a_{23} & a_{21} & a_{22} = a_{31}a_{22}a_{13} + a_{32}a_{23}a_{11} + a_{33}a_{21}a_{12} \\
a_{31} & a_{32} & a_{33} & a_{31} & a_{32}
\end{array}
$$

from the products of the diagonal terms (multiply terms from upper left to lower right)

$$
\begin{array}{ccccc}
a_{11} & a_{12} & a_{13} & a_{11} & a_{12} \\[2pt]
& \ddots & \ddots & \ddots & \\[2pt]
a_{21} & a_{22} & a_{23} & a_{21} & a_{22} = a_{11}a_{22}a_{33} + a_{12}a_{23}a_{31} + a_{13}a_{21}a_{32} \\[2pt]
& \ddots & \ddots & \ddots & \\[2pt]
a_{31} & a_{32} & a_{33} & a_{31} & a_{32}
\end{array}
$$

EXERCISE 4.9: Evaluate

$$
\det \begin{bmatrix} 2 & 4 & 3 \\ 6 & 1 & 5 \\ -2 & 1 & 3 \end{bmatrix}
$$

In general, the determinant of a square matrix can be calculated in terms of the cofactor. The cofactor of the square matrix A is denoted by $\mathrm{cof}_{ij}(A)$ and has the definition:

$\mathrm{cof}_{ij}(A) \equiv$ determinant of A found by deleting the ith row and j^{th} column of A and choosing the positive (negative) sign if $i + j$ is even (odd).

The integer $i + j$ is even if it is part of the sequence of integers $\{2, 4, 6, \ldots\}$, otherwise $i + j$ is odd.

EXERCISE 4.10: Use the definition of cofactor to evaluate

$$
\mathrm{cof}_{23} \begin{bmatrix} 2 & 4 & 3 \\ 6 & 1 & 5 \\ -2 & 1 & 3 \end{bmatrix}
$$

The determinant of an arbitrary square matrix A of order $n \times n$ can be evaluated using cofactors; thus

$$
|A| = \sum_{j=1}^{n} a_{ij} \mathrm{cof}_{ij}(A) \text{ for any } i
$$

or

$$|A| = \sum_{i=1}^{n} a_{ij}\text{cof}_{ij}(A) \text{ for any } j$$

These expressions represent an expansion by cofactors.

EXERCISE 4.11: Evaluate the following determinant using cofactors:

$$|A| = \det \begin{bmatrix} 2 & 4 & 3 \\ 6 & 1 & 5 \\ -2 & 1 & 3 \end{bmatrix}$$

Determinants are useful for determining whether or not a square matrix A is singular or nonsingular. If $\det(A) \neq 0$, then there exists a unique inverse matrix A^{-1}. Furthermore, if square matrices A and B are nonsingular, then

$$(AB)^{-1} = B^{-1}A^{-1}$$

Application: Cramer's Rule. Suppose that a linear system of equations can be written in terms of the nonsingular matrix A, the unknown column vector x, and a known column vector b such that

$$Ax = b \tag{4.3.1}$$

Recall that $\det(A) \neq 0$ for a nonsingular matrix A. Cramer's rule says that Eq. (4.3.1) has the unique solution

$$x_j = \frac{\det(A_j)}{\det(A)} \tag{4.3.2}$$

where A_j is a matrix obtained from A by replacing the jth column of A by b. For example, suppose A is a 2×2 square matrix and b is a 2×1 column matrix; that is,

$$A = \begin{bmatrix} a_{11} & a_{12} \\ a_{21} & a_{22} \end{bmatrix}, b = \begin{bmatrix} b_1 \\ b_2 \end{bmatrix}$$

In this case, there are two solutions of Eq. (4.3.2), namely,

$$x_1 = \frac{\det \begin{bmatrix} b_1 & a_{12} \\ b_2 & a_{22} \end{bmatrix}}{\det(A)}, x_2 = \frac{\det \begin{bmatrix} a_{11} & b_1 \\ a_{21} & b_2 \end{bmatrix}}{\det(A)}$$

If A is singular, that is, $\det(A) = 0$, then Eq. (4.3.1) has either no solution or an infinite number of solutions.

EXERCISE 4.12: Use Cramer's rule to find the unknowns $\{x_1, x_2, x_3\}$ that satisfy the system of equations

$$x_1 + 2x_3 = 11$$
$$2x_1 - x_2 + 3x_3 = 17$$
$$4x_1 + x_2 + 8x_3 = 45$$

CHAPTER 5

LINEAR ALGEBRA II

Matrices arise naturally in the treatment of systems of linear algebraic equations with the form $Ax = B$, where A is an $m \times n$ matrix and x, B are $n \times 1$ matrices. Another perspective is to view x and B as vectors and A as an operator that transforms x into B. We review the mathematical framework associated with the vector perspective in this chapter and show how to calculate eigenvalues and eigenvectors. An important practical application of this material is to the diagonalization of a matrix, which is presented in an application.

5.1 VECTORS

A vector A from the origin of Figure 5.1 to a point P in three dimensions has the form

$$A = a_x\hat{i} + a_y\hat{j} + a_z\hat{k} \tag{5.1.1}$$

where $\{\hat{i}, \hat{j}, \hat{k}\}$ are unit vectors along the $\{x, y, z\}$ axes, respectively, and the vector components $\{a_x, a_y, a_z\}$ are the corresponding distances along the $\{x, y, z\}$ axes. The length or magnitude of vector A is

$$|A| = \sqrt{a_x^2 + a_y^2 + a_z^2} \tag{5.1.2}$$

Math Refresher for Scientists and Engineers, Third Edition By John R. Fanchi
Copyright © 2006 John Wiley & Sons, Inc.

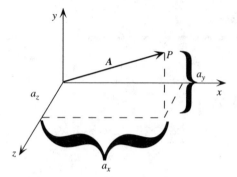

Figure 5.1 Vector components.

A unit vector is a vector with magnitude one. The vector A may be written as an ordered triplet such that

$$A = (a_x, a_y, a_z)$$

where the elements of the ordered triplet are the vector components.

Example: The unit vectors are

$$\hat{i} = (1, 0, 0)$$

$$\hat{j} = (0, 1, 0)$$

$$\hat{k} = (0, 0, 1)$$

Addition and subtraction of two vectors A and B are given by

$$A \pm B = (a_x \pm b_x)\hat{i} + (a_y \pm b_y)\hat{j} + (a_z \pm b_z)\hat{k} \qquad (5.1.3)$$

Multiplying a vector A by a scalar α gives

$$\alpha A = \alpha\, a_x \hat{i} + \alpha\, a_y \hat{j} + \alpha\, a_z \hat{k}$$

A zero vector is defined as

$$a_x \hat{i} + a_y \hat{j} + a_z \hat{k} = 0$$

which implies that $a_x = a_y = a_z = 0$.

Example: Let $B = (3, 2, 0)$ and $C = (3, -2, 0)$. Then $|B| = |C|$ but the orientation of B does not equal the orientation of C.

The dot product of two vectors is

$$A \cdot B = a_x b_x + a_y b_y + a_z b_z \qquad (5.1.4)$$

or

$$A \cdot B = |A| \, |B| \cos \theta$$

where the angle θ is the angle between the vectors A, B as in Figure 5.2. Notice that the dot product is a maximum when the vectors are aligned ($\theta = 0$), a minimum when the vectors point in opposite directions ($\theta = 180°$), and zero when the vectors are perpendicular ($\theta = 90°$).

The dot product $A \cdot B$ is commutative; that is,

$$A \cdot B = B \cdot A$$

The dot product of a vector with itself gives the square of the magnitude of the vector.

EXERCISE 5.1: Define $C = B - A$. Derive the law of cosines by calculating $C \cdot C$ for $\theta \leq 90°$.

Vectors may be written as matrices. For example, in matrix notation, the vectors $A = (a_x, a_y, a_z)$ and $B = (b_x, b_y, b_z)$ are the 3×1 matrices

$$A = \begin{bmatrix} a_x \\ a_y \\ a_z \end{bmatrix}, B = \begin{bmatrix} b_x \\ b_y \\ b_z \end{bmatrix}$$

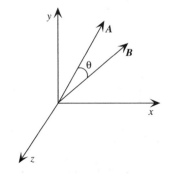

Figure 5.2   Angle between vectors.

The dot product in matrix notation is

$$A \cdot B = A^T B = [a_x \, a_y \, a_z] \begin{bmatrix} b_x \\ b_y \\ b_z \end{bmatrix} = a_x b_x + a_y b_y + a_z b_z \qquad (5.1.5)$$

The dot product is an operation between two vectors that yields a scalar. An operation between two vectors that yields a vector is the *cross product*, which has the definition

$$C = A \times B = \hat{n} |A| \, |B| \sin \theta = -B \times A \qquad (5.1.6)$$

where the vector C and the unit vector $\hat{n}$ are normal (perpendicular) to the plane containing both vectors A and B. The unit vector $\hat{n}$ is parallel to the vector C in Figure 5.3. Both $\hat{n}$ and C are oriented in the direction of advance of a right-handed screw if the screw was rotated through the angle θ shown in Figure 5.3 [Kreyszig, 1999]. In general, the cross product $A \times B$ is antisymmetric; that is, $A \times B = -B \times A$.

Example: The cross product of each pair of unit vectors gives

$$\hat{i} \times \hat{j} = -\hat{j} \times \hat{i} = \hat{k}$$

$$\hat{j} \times \hat{k} = -\hat{k} \times \hat{j} = \hat{i}$$

$$\hat{k} \times \hat{i} = -\hat{i} \times \hat{k} = \hat{j}$$

An alternative expression for the cross product in terms of Cartesian unit vectors is

$$A \times B = \det \begin{bmatrix} \hat{i} & \hat{j} & \hat{k} \\ a_x & a_y & a_z \\ b_x & b_y & b_z \end{bmatrix}$$

$$= (a_y b_z - a_z b_y)\hat{i} + (a_z b_x - a_x b_z)\hat{j} + (a_x b_y - a_y b_x)\hat{k}$$

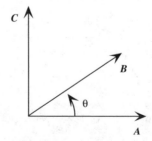

Figure 5.3 Orientation of vectors in cross product.

Example: Let $A = (3, 2, 0)$ and $B = (3, -2, 0)$. Then

$$A \times B = \det \begin{bmatrix} \hat{i} & \hat{j} & \hat{k} \\ 3 & 2 & 0 \\ 3 & -2 & 0 \end{bmatrix} = (-6 - 6)\hat{k} = -12\hat{k}$$

Notice that vectors A and B are in the x, y plane, but $A \times B$ projects in the negative z direction.

5.2 VECTOR SPACES

The idea of associating a column matrix with a vector leads to the concept of vector spaces. Let E^n denote the set of all $n \times 1$ column matrices, or vectors with n elements. The nonempty set E^n is a vector space because it satisfies the following axioms for vectors $u, v, w \in E^n$ and scalars a, b.

Closure		$u + v \in E^n, au \in E^n$
Addition	Associative law	$(u + v) + w = u + (v + w)$
	Zero vector	$u + 0 = u$
	Existence of negatives	$u + (-u) = 0$
	Commutative law	$u + v = v + u$
Multiplication	Distributive law	$a(u + v) = au + av$
		$(a + b)u = au + bu$
	Associative law	$(ab)u = a(bu)$
	Unity law	$1u = u$

A set of vectors $\{u^{(1)}, u^{(2)}, \dots, u^{(n)}\}$ is linearly dependent if and only if the sum

$$\sum_{i=1}^{n} c_i u^{(i)} = 0 \tag{5.2.1}$$

is satisfied for a set of scalars $\{c_1, \dots, c_m\}$ that are not all zero. The set of vectors $\{u^{(1)}, u^{(2)}, \dots, u^{(n)}\}$ is linearly independent if Eq. (5.2.1) is satisfied only when $c_i = 0$ for all values of i.

A basis for the set E^n is a set of n linearly independent vectors that belong to E^n. If we write the basis as $\{u^{(1)}, u^{(2)}, \dots, u^{(n)}\}$, then any vector $\mathbf{u} \in E^n$ can be written as a linear combination, or superposition, of basis vectors such that

$$u = \sum_{i=1}^{n} c_i u^{(i)}$$

where $\{c_i\}$ is a unique set of numbers.

The inner product of two n-dimensional vectors u, v is defined as

$$(u, v) = \sum_{i=1}^{n} u_i^* v_i = (u^*)^T v = u \cdot v$$

where $*$ denotes complex conjugation. Two vectors $u^{(i)}, u^{(j)}$ are orthogonal if their inner product is zero; that is, $(u^{(i)}, u^{(j)}) = 0$. Vectors $u^{(i)}, u^{(j)}$ are orthonormal if their inner product satisfies

$$(u^{(i)}, u^{(j)}) = \delta_{ij}$$

where the Kronecker delta is defined by

$$\delta_{ij} = \left\{ \begin{array}{ll} 0, & i \neq j \\ 1, & i = j \end{array} \right\}$$

The Euclidean length of a vector u with elements $\{u_i : i = 1, \ldots, n\}$ is the inner product of vector u with itself; thus

$$\|u\| = (u, u)^{1/2} = \left[\sum_{i=1}^{n} |u_i|^2 \right]^{1/2}$$

The Euclidean lengths of two vectors u, v satisfy the Schwartz inequality

$$|u \cdot v| \leq \|u\| \, \|v\|$$

Example: A set of basis vectors in E^3 is

$$e_1 = \begin{bmatrix} 1 \\ 0 \\ 0 \end{bmatrix}, e_2 = \begin{bmatrix} 0 \\ 1 \\ 0 \end{bmatrix}, e_3 = \begin{bmatrix} 0 \\ 0 \\ 1 \end{bmatrix}$$

The vector $B = (3, 2, 0)$ may be written in terms of basis vectors as

$$B = 3e_1 + 2e_2 + 0e_3$$

with the coefficients $\{c_i\} = \{3, 2, 0\}$. Vector B has the Euclidean length

$$\|B\| = (3^2 + 2^2)^{1/2} = \sqrt{13}$$

5.3 EIGENVALUES AND EIGENVECTORS

Suppose A is an $n \times n$ matrix. Then the determinant

$$|A - \lambda I| = 0 \tag{5.3.1}$$

is the characteristic equation of A and I is the identity matrix. The characteristic equation is an nth degree polynomial resulting from expansion of the determinant. The n values of λ are the characteristic roots of the characteristic equation. They are also the eigenvalues associated with matrix A.

EXERCISE 5.2: Let

$$A = \begin{bmatrix} a_{11} & a_{12} \\ a_{21} & a_{22} \end{bmatrix}$$

Find the characteristic equation and the characteristic roots.

Eigenvalues are calculated from the eigenvalue equation

$$Ax = \lambda x$$

where x is a column vector with n rows and λ is the eigenvalue of A. The eigenvalue equation may be written in the form

$$Ax = \lambda x = \lambda I x$$

or

$$(A - \lambda I)x = 0 \tag{5.3.2}$$

Equation (5.3.2) has nonzero solutions if λ is a characteristic root of A; that is, λ is a solution of Eq. (5.3.1). The solution x is the characteristic vector. If x has unit length; that is,

$$x^T x = 1$$

it becomes the eigenvector for the characteristic root λ of A.

The roots or eigenvalues depend on the properties of the matrix A. For example, a real symmetric matrix has real roots, and a Hermitian matrix has real roots. If A is a square matrix of order n with eigenvalues $\{\lambda_1, \lambda_2, \ldots, \lambda_n\}$, then the set of n eigenvalues is called the *spectrum* of A. The spectral radius of A is the eigenvalue with the largest magnitude. The trace and the determinant of A are given in terms of eigenvalues as

$$\det(A) = \prod_{i=1}^{n} \lambda_i, \ \mathrm{Tr}\,(A) = \sum_{i=1}^{n} \lambda_i$$

If matrices A, B are similar, then they have the same eigenvalues.

EXERCISE 5.3: Find eigenvalues for the Pauli matrices given in Exercise 4.1; that is, find λ from

$$\boldsymbol{\sigma}_i \psi = \lambda I \psi$$

where ψ is an eigenfunction and $i = \{x, y, z\}$.

Hamilton–Cayley Theorem

In general, the characteristic equation associated with a square matrix A of order n may be written as the polynomial

$$f(\lambda) = \sum_{i=0}^{n} a_i \lambda^{n-i} = 0$$

where λ are the eigenvalues given by the characteristic determinant

$$|A - \lambda I| = 0$$

If we replace λ in $f(\lambda)$ by the matrix A so that we obtain

$$f(A) = \sum_{i=0}^{n} a_i A^{n-i}$$

then the Hamilton–Cayley theorem says that

$$f(A) = \mathbf{0}$$

In other words, the matrix A satisfies its characteristic equation; thus

$$\sum_{i=0}^{n} a_i A^{n-i} = \mathbf{0}$$

Example: We demonstrate the validity of the Hamilton–Cayley theorem by applying it to the matrix

$$A = \begin{bmatrix} 5 & 4 \\ 1 & 2 \end{bmatrix}$$

The characteristic equation is

$$\begin{vmatrix} 5-\lambda & 4 \\ 1 & 2-\lambda \end{vmatrix} = (5-\lambda)(2-\lambda) - 4 = \lambda^2 - 7\lambda + 6 = 0$$

so that the eigenvalues are

$$\lambda_{\pm} = \frac{7 \pm \sqrt{49-24}}{2} = \frac{7 \pm 5}{2}$$

In polynomial form,

$$f(\lambda) = \lambda^2 - 7\lambda + 6 = 0$$

so that

$$f(A) = A^2 - 7A + 6I$$

Substituting in the explicit representation of A gives the expected result

$$f(A) = \begin{bmatrix} 5 & 4 \\ 1 & 2 \end{bmatrix}\begin{bmatrix} 5 & 4 \\ 1 & 2 \end{bmatrix} - 7\begin{bmatrix} 5 & 4 \\ 1 & 2 \end{bmatrix} + \begin{bmatrix} 6 & 0 \\ 0 & 6 \end{bmatrix}$$

$$= \begin{bmatrix} 29 & 28 \\ 7 & 8 \end{bmatrix} - \begin{bmatrix} 35 & 28 \\ 7 & 14 \end{bmatrix} + \begin{bmatrix} 6 & 0 \\ 0 & 6 \end{bmatrix} = \begin{bmatrix} 0 & 0 \\ 0 & 0 \end{bmatrix}$$

The Hamilton–Cayley theorem is useful in evaluating the inverse of a square matrix. To show this, we formally multiply $f(A)$ by A^{-1} to find

$$A^{-1}f(A) = a_0 A^{n-1} + a_1 A^{n-2} + \cdots + a_{n-1}I + a_n A^{-1} = 0$$

Solving for A^{-1} gives

$$A^{-1} = -\frac{1}{a_n}\left[\sum_{i=0}^{n-1} a_i A^{n-1-i} \right]$$

The inverse of a square matrix can be found from its characteristic equation and by forming the powers of A up to $n-1$.

Example: Use the Hamilton–Cayley theorem to determine the inverse of

$$A = \begin{bmatrix} 5 & 4 \\ 1 & 2 \end{bmatrix}$$

From the above example we know that the characteristic equation is

$$\lambda^2 - 7\lambda + 6 = 0$$

so that $a_0 = 1$, $a_1 = -7$, $a_2 = 6$, and $n = 2$. Thus

$$A^{-1} = -\frac{1}{a_2}\left[\sum_{i=0}^{1} a_i A^{2-1-i}\right] = -\frac{1}{a_2}[a_0 A + a_1 I]$$

Substituting in the explicit representation of A gives

$$A^{-1} = -\frac{1}{6}\left\{\begin{bmatrix} 5 & 4 \\ 1 & 2 \end{bmatrix} - 7\begin{bmatrix} 1 & 0 \\ 0 & 1 \end{bmatrix}\right\}$$

Simplifying gives

$$A^{-1} = -\frac{1}{6}\left\{\begin{bmatrix} 5 & 4 \\ 1 & 2 \end{bmatrix} - \begin{bmatrix} 7 & 0 \\ 0 & 7 \end{bmatrix}\right\} = -\frac{1}{6}\begin{bmatrix} -2 & 4 \\ 1 & -5 \end{bmatrix}$$

or

$$A^{-1} = \begin{bmatrix} \dfrac{2}{6} & -\dfrac{4}{6} \\ -\dfrac{1}{6} & \dfrac{5}{6} \end{bmatrix}$$

To verify, we evaluate the matrix product

$$AA^{-1} = \begin{bmatrix} 5 & 4 \\ 1 & 2 \end{bmatrix}\begin{bmatrix} \dfrac{2}{6} & -\dfrac{4}{6} \\ -\dfrac{1}{6} & \dfrac{5}{6} \end{bmatrix} = \begin{bmatrix} \dfrac{10}{6} - \dfrac{4}{6} & -\dfrac{20}{6} + \dfrac{20}{6} \\ \dfrac{2}{6} - \dfrac{2}{6} & -\dfrac{4}{6} + \dfrac{10}{6} \end{bmatrix}$$

$$= \begin{bmatrix} 1 & 0 \\ 0 & 1 \end{bmatrix} = I$$

which is the expected result.

EXERCISE 5.4: Use the Hamilton–Cayley theorem to find the inverse of the complex Pauli matrix

$$\sigma_y = \begin{bmatrix} 0 & -i \\ i & 0 \end{bmatrix}$$

5.4   MATRIX DIAGONALIZATION

It is often worthwhile to transform a matrix representation from a coordinate system x in which an $n \times n$ square matrix D has nonzero off-diagonal elements to a coordinate system y where only the diagonal elements of a square matrix D' are nonzero. This can be done by applying a similarity transformation

$$D' = ADA^{-1} \qquad (5.4.1)$$

to the matrix D, where A is an invertible matrix. The coordinate systems x and y are related by the transformation matrix A such that

$$y = Ax \qquad (5.4.2)$$

The procedure for finding A is the subject of this section.

Similarity Transformations

Suppose two matrices α and β are related by the linear transformation

$$\alpha = D\beta \qquad (5.4.3)$$

Such a relationship exists in many applications. For example, angular momentum is related to angular velocity by a linear transformation [Goldstein et al., 2002]. The transformation matrix D in this example is the moment of inertia. As noted previously, there are occasions when it is useful to diagonalize D while preserving the form of the relationship between α and β.

It is easy to show that a similarity transformation preserves the relationship in Eq. (5.4.3). First, multiply Eq. (5.4.3) on the left by A to find

$$A\alpha = AD\beta \qquad (5.4.4)$$

where A is a nonsingular, $n \times n$ square matrix. Since A is nonsingular, it is invertible; that is, it satisfies the equality

$$A^{-1}A = AA^{-1} = I \qquad (5.4.5)$$

where I is the $n \times n$ square matrix. Substituting Eq. (5.4.5) into Eq. (5.4.4) gives

$$A\alpha = ADA^{-1}A\beta \qquad (5.4.6)$$

Defining the transformed matrices

$$\begin{aligned} \alpha' &= A\alpha \\ \beta' &= A\beta \end{aligned} \qquad (5.4.7)$$

and using the similarity transformation

$$D' = ADA^{-1} \qquad (5.4.8)$$

in Eq. (5.4.6) yields

$$\alpha' = D'\beta' \qquad (5.4.9)$$

Equation (5.4.9) is the same form as Eq. (5.4.3).

Diagonalization Algorithm

The matrix D is diagonalized by finding and applying a particular similarity transformation matrix A. The procedure for finding a matrix A that diagonalizes D is given by Nicholson [1986] and summarized as follows:

Algorithm for Diagonalizing a Square Matrix

Given an $n \times n$ matrix D:

1. Find the eigenvalues $\{\lambda_i : i = 1, \ldots, n\}$ of D.
2. Find n linearly independent eigenvectors $\{a_i : i = 1, \ldots, n\}$.
3. Form the similarity transformation matrix A with the eigenvectors as columns.
4. Calculate the diagonalized similar matrix D'. The diagonal entries of D' are the eigenvalues corresponding to the eigenvectors $\{a_i : i = 1, \ldots, n\}$

The next application illustrates the diagonalization algorithm.

Application: Diagonalizing a Square Matrix. The ideas just presented are made more concrete by applying the matrix diagonalization algorithm to a specific example. Consider the 2×2 matrix

$$D = \begin{bmatrix} d_{11} & \epsilon_{12} \\ \epsilon_{21} & d_{22} \end{bmatrix} \tag{5.4.10}$$

as viewed in coordinate system x. For this example, we arbitrarily require that the elements of D satisfy the relations [Fanchi, 1983]

$$d_{11} \neq d_{22}$$
$$\epsilon_{12} = \epsilon_{21} \tag{5.4.11}$$

To find the diagonal matrix D' corresponding to D we must first solve the eigenvalue problem

$$\det[D - \lambda I] = 0 \tag{5.4.12}$$

EXERCISE 5.5: Calculate the eigenvalues of the matrix D.

The characteristic roots or eigenvalues $\{\lambda\}$ of Eq. (5.4.12) are the diagonal elements of the diagonalized matrix D'; thus

$$D' = \begin{bmatrix} \lambda_+ & 0 \\ 0 & \lambda_- \end{bmatrix} \tag{5.4.13}$$

where λ_+, λ_- were calculated in Exercise 5.5.

Eigenvectors

The matrix A is composed of orthonormal eigenvectors $\{a\}$ found from

$$Da = \lambda a \qquad (5.4.14)$$

The basis vector a satisfies

$$(D - \lambda I)a = 0 \qquad (5.4.15)$$

where I is the identity matrix. Expanding Eq. (5.4.15) gives

$$
\begin{aligned}
(d_{11} - \lambda_+)a_1^+ + \epsilon_{12}a_2^+ &= 0 \\
\epsilon_{12}a_1^+ + (d_{22} - \lambda_+)a_2^+ &= 0
\end{aligned}
\qquad (5.4.16)
$$

for the eigenvalue λ_+, and

$$
\begin{aligned}
(d_{11} - \lambda_-)a_1^- + \epsilon_{12}a_2^- &= 0 \\
\epsilon_{12}a_1^- + (d_{22} - \lambda_-)a_2^- &= 0
\end{aligned}
$$

for the eigenvalue λ_-. Rearranging Eq. (5.4.16) gives

$$a_1^+ = \frac{-\epsilon_{12}}{d_{11} - \lambda_+} a_2^+ \qquad (5.4.17)$$

Equation (5.4.17) and the normalization condition

$$(a_1^+)^2 + (a_2^+)^2 = 1 \qquad (5.4.18)$$

provide the two equations that are necessary for determining the components of a^+; thus

$$a_2^+ = \left\{ 1 + \frac{\epsilon_{12}^2}{(d_{11} - \lambda_+)^2} \right\}^{-1/2}$$

and

$$a_1^+ = \frac{-\epsilon_{12}}{(d_{11} - \lambda_+)} \left\{ 1 + \frac{\epsilon_{12}^2}{(d_{11} - \lambda_+)^2} \right\}^{-1/2}$$

Similar calculations for a^- yield the results

$$a_1^- = \frac{-(d_{11} - \lambda_+)}{\{(d_{11} - \lambda_+)^2 + \epsilon_{12}^2\}^{1/2}}$$

and

$$a_2^- = \frac{-\epsilon_{12}}{\{(d_{11} - \lambda_+)^2 + \epsilon_{12}^2\}^{1/2}}$$

where the relation

$$d_{11} - \lambda_+ = -(d_{22} - \lambda_-)$$

has been used.

EXERCISE 5.6: Show that a^+ and a^- are orthogonal.

Coordinate Transformation

We now use the orthonormal eigenvectors a^+ and a^- to construct the transformation matrix A. According to the algorithm for diagonalizing a square matrix presented previously, we form A as

$$A = [a^+, a^-] = \begin{bmatrix} a_1^+ & a_1^- \\ a_2^+ & a_2^- \end{bmatrix}$$

or

$$A = \frac{1}{[(d_{11} - \lambda_+)^2 + \epsilon_{12}^2]^{1/2}} \begin{bmatrix} -\epsilon_{12} & -(d_{11} - \lambda_+) \\ d_{11} - \lambda_+ & -\epsilon_{12} \end{bmatrix} \qquad (5.4.19)$$

The transformed coordinate system y is given by $y = Ax$. Rewriting the matrix equation for coordinate transformations in algebraic form gives

$$y_1 = a_1^+ x_1 + a_1^- x_2$$
$$y_2 = a_2^+ x_1 + a_2^- x_2$$

or

$$\begin{bmatrix} y_1 \\ y_2 \end{bmatrix} = \begin{bmatrix} a_1^+ & a_1^- \\ a_2^+ & a_2^- \end{bmatrix} \begin{bmatrix} x_1 \\ x_2 \end{bmatrix} \qquad (5.4.20)$$

An angle θ is associated with the linear transformation by noting from Section 4.1 that

$$\begin{bmatrix} y_1 \\ y_2 \end{bmatrix} = \begin{bmatrix} \cos\theta & \sin\theta \\ -\sin\theta & \cos\theta \end{bmatrix} \begin{bmatrix} x_1 \\ x_2 \end{bmatrix} \qquad (5.4.21)$$

Equating elements of the transformation matrix in Eqs. (5.4.20) and (5.4.21) gives

$$a_1^+ = a_2^- = \cos\theta$$

and

$$a_2^+ = -a_1^- = -\sin\theta$$

EXERCISE 5.7: Verify that $a_1^+ = a_2^-$ and $a_2^+ = -a_1^-$. Find θ.

CHAPTER 6

DIFFERENTIAL CALCULUS

Calculus provides a framework for studying change which has widespread practical utility. We begin a review of this framework by presenting the concept of limits and extending the concept to the calculation of derivatives.

6.1 LIMITS

Let $f(x)$ be a real-valued function of a real variable x and let a be some fixed point in the domain of f. The number L is the limit of $f(x)$ as x approaches a if, given any positive number ε, we can find a positive number δ such that $f(x)$ is in the range

$$|f(x) - L| < \varepsilon$$

whenever x is in the domain

$$0 < |x - a| < \delta$$

We write

$$\lim_{x \to a} f(x) = L$$

and say "the limit of $f(x)$ as x approaches a is L."

Math Refresher for Scientists and Engineers, Third Edition By John R. Fanchi
Copyright © 2006 John Wiley & Sons, Inc.

Limits may be evaluated graphically or analytically. One straightforward method for evaluating a limit is to first evaluate $f(x)$ at $f(a + \varepsilon)$ and $f(a - \varepsilon)$, then let $\varepsilon \to 0$. If $f(a + \varepsilon) = f(a - \varepsilon)$ when $\varepsilon \to 0$, then the limit exists and is equal to $f(a)$.

Example

$$\lim_{x \to 1} 4x - 2 = 2$$

$$\lim_{x \to -3} x^2 + 3 = 12$$

Some limits are more difficult to find if they contain a denominator that goes to zero as $x \to a$. These limits must be treated carefully using, for example, L'Hôpital's rule, discussed in Section 6.2.

Proof that a limit exists requires manipulation of inequalities to determine the relationship between ε and δ. It is also necessary to verify that the inequalities $|f(x) - L| < \varepsilon$ and $0 < |x - a| < \delta$ are satisfied as required by the definition of a limit. Examples of proofs of limits can be found in standard calculus textbooks such as Stewart [1999], Larson et al. [1990], and Thomas [1972]. We now consider several useful properties of limits.

Suppose we have the limits

$$\lim_{x \to a} f(x) = L \tag{6.1.1}$$

and

$$\lim_{x \to a} g(x) = M \tag{6.1.2}$$

then

$$\lim_{x \to a} [f(x) + g(x)] = L + M$$

that is, the limit of the sum of two functions is the sum of the limits.

EXERCISE 6.1: Let $f(x) = 3x + 7$, $g(x) = 5x + 1$, so that $f(x) + g(x) = 8x + 8$. Find $\lim_{x \to 2} [f(x) + g(x)]$.

The limit of a constant c times a function $f(x)$ is the constant c times the limit L of the function; that is,

$$\lim_{x \to a} c f(x) = c[\lim_{x \to a} f(x)] = cL$$

where we have used Eq. (6.1.1).

EXERCISE 6.2: Let $f(x) = 3x + 7$ and $c = 6$. Find $\lim_{x \to 2} 6(3x + 7)$.

The limit of the ratio of two functions $f(x)$, $g(x)$ is the ratio of their respective limits if the limit of the denominator is not 0; that is,

$$\lim_{x \to a} \frac{f(x)}{g(x)} = \frac{L}{M} \text{ for } M \neq 0$$

where we have used Eqs. (6.1.1) and (6.1.2).

EXERCISE 6.3: Find

$$\lim_{x \to 2} \frac{3x + 7}{5x + 1}$$

If $P(x)$ is a polynomial of the form

$$P(x) = \sum_{i=0}^{n} \alpha_i x^{n-i}$$

with constant coefficients $\{\alpha_i\}$, then

$$\lim_{x \to a} P(x) = P(a)$$

EXERCISE 6.4: Let $P(x) = \alpha x^2 + \beta x + \gamma$. Find $\lim_{x \to 0} P(x)$.

If $f(x)$ is any rational function; that is,

$$f(x) = \frac{N(x)}{D(x)}$$

where $N(x)$, $D(x)$ are polynomials, and $D(a) \neq 0$, then

$$\lim_{x \to a} f(x) = \frac{N(a)}{D(a)}$$

EXERCISE 6.5: Find

$$\lim_{x \to -1} \frac{x^3 + 5x^2 + 3x}{2x^5 + 3x^4 - 2x^2 - 1}$$

6.2 DERIVATIVES

The definition of the derivative of a function $f(x)$ with respect to the independent variable x is given in terms of the limit of the function evaluated at two points x and $x + \Delta x$; thus

$$\frac{df(x)}{dx} = f'(x) = \lim_{\Delta x \to 0} \frac{f(x + \Delta x) - f(x)}{\Delta x} \tag{6.2.1}$$

The derivative, which is often written $f'(x)$, is the slope of the line that is tangent to the graph of $f(x)$ at the point x. Slope is illustrated by the dashed line in Figure 6.1. Notice that, in the limit as Δx approaches 0, the dashed line in Figure 6.1 more accurately approximates the tangent to the graph of $f(x)$ at the point x. Derivatives for several common functions are derived from the definition of a derivative in the following exercise.

EXERCISE 6.6: Use the definition of the derivative to find the derivative of each of the following functions: (a) $f(x) = ax$; (b) $f(x) = bx^2$; (c) $f(x) = e^x$; (d) $f(x) = \sin x$.

Several common derivatives are tabulated as follows:

DERIVATIVES

Assume $\{u, v\}$ are functions of x, and $\{a, n\}$ are constants.

$$\frac{d}{dx} a = 0 \qquad\qquad \frac{d}{dx}(u + v) = \frac{du}{dx} + \frac{dv}{dx}$$

$$\frac{d}{dx} au = a\frac{du}{dx} \qquad\qquad \frac{d}{dx}(uv) = u\frac{dv}{dx} + v\frac{du}{dx}$$

$$\frac{d}{dx} u^n = nu^{n-1}\frac{du}{dx} \qquad\qquad \frac{d}{dx}\left(\frac{u}{v}\right) = \frac{1}{v}\frac{du}{dx} - \frac{u}{v^2}\frac{dv}{dx}$$

$$\frac{d}{dx}\sin u = \cos u\frac{du}{dx} \qquad\qquad \frac{d}{dx}\sinh u = \cosh u\frac{du}{dx}$$

$$\frac{d}{dx}\cos u = -\sin u\frac{du}{dx} \qquad\qquad \frac{d}{dx}\cosh u = \sinh u\frac{du}{dx}$$

$$\frac{d}{dx}\tan u = \sec^2 u\frac{du}{dx} \qquad\qquad \frac{d}{dx}\tanh u = \text{sech}^2 u\frac{du}{dx}$$

$$\frac{d}{dx} e^u = e^u\frac{du}{dx} \qquad\qquad \frac{d}{dx}\ln u = \frac{1}{u}\frac{du}{dx}$$

$$\frac{d}{dx} a^u = a^u \ln a\frac{du}{dx} \qquad\qquad \frac{d}{dx}\log_a u = \frac{\log_a e}{u}\frac{du}{dx};$$

$$a > 0, \ a \neq 1$$

$$\frac{d}{dx}\sin^{-1}u = \frac{1}{\sqrt{1-u^2}}\frac{du}{dx} \qquad \frac{d}{dx}\sinh^{-1}u = \frac{1}{\sqrt{1+u^2}}\frac{du}{dx}$$

$$\frac{d}{dx}\cos^{-1}u = -\frac{1}{\sqrt{1-u^2}}\frac{du}{dx} \qquad \frac{d}{dx}\cosh^{-1}u = \frac{1}{\sqrt{u^2-1}}\frac{du}{dx}$$

$$\frac{d}{dx}\tan^{-1}u = \frac{1}{1+u^2}\frac{du}{dx} \qquad \frac{d}{dx}\tanh^{-1}u = \frac{1}{1-u^2}\frac{du}{dx}; \ |u| < 1$$

L'Hôpital's Rule

Let $f(x), g(x)$ be two continuous functions in the interval $a \le x \le b$ and be differentiable in the interval $a < x < b$. Suppose $f(a) = g(a) = 0$, as in Figure 6.2. If the derivative $f'(x) \ne 0$ in the interval $a < x < b$, L'Hôpital's rule says that

$$\lim_{x \to a^+} \frac{g(x)}{f(x)} = \lim_{x \to a^+} \frac{g'(x)}{f'(x)} \tag{6.2.2}$$

whenever the right-hand side exists. The notation $x \to a^+$ signifies that x approaches a from above, which means we start with values of $x > a$ and then reduce x to approach a.

Example: For the curves shown in Figure 6.2, an application of L'Hôpital's rule yields a limit of 0 as x approaches a from above. This can be verified by observing that $g(x)$ has 0 slope at $x = a$, while $f(x)$ has a positive, nonzero slope at $x = a$. Thus $g'(x) = 0$ at $x = a$ so that $g'(x)/f'(x) = 0$ at $x = a$ since $f'(x) > 0$ at $x = a$.

EXERCISE 6.7: Evaluate

$$\lim_{x \to 0} \frac{x^2 + bx}{x}$$

where $g(x) = x^2 + bx$, and $f(x) = x$.

In some cases, it is difficult to calculate the limit, even using L'Hôpital's rule. Exercise 6.8 illustrates this point.

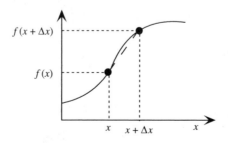

Figure 6.1 Illustration of slope.

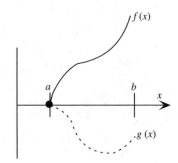

Figure 6.2   Illustration of continuous functions for L'Hôpital's rule.

EXERCISE 6.8: Evaluate

$$\lim_{x \to 0} \frac{x}{\ln^2 (1 + x)}$$

where $g(x) = x$, and $f(x) = \ln^2 (1 + x)$.

In general, L'Hôpital's rule applies to limits of indeterminate forms: $0/0, \infty/\infty, (-\infty)/\infty, \infty/(-\infty)$, and $(-\infty)/(-\infty)$. For further details and examples, see Larson et al. [1990].

Properties of Derivatives

The derivative of a product of two functions $u(x)$, $v(x)$ is

$$\frac{d}{dx} uv = u\frac{dv}{dx} + v\frac{du}{dx}$$

The derivative of the sum of two functions is the sum of their respective derivatives; thus

$$\frac{d}{dx}(u + v) = \frac{du}{dx} + \frac{dv}{dx}$$

Suppose we let c be a constant. The derivative of a constant is zero:

$$\frac{dc}{dx} = 0$$

The following derivatives can be derived from the preceding relations for constants c, n and functions $u(x)$, $v(x)$, $w(x)$:

$$\frac{dcv}{dx} = c\frac{dv}{dx}$$

$$\frac{du^n}{dx} = nu^{n-1}\frac{du}{dx} \quad \text{for } n \neq 0$$

If $n = 0$, the derivative of u^0 equals 0 because $u^0 = 1$ is a constant. The derivative of the quotient u/w is

$$\frac{d}{dx}(u/w) = \frac{1}{w}\frac{du}{dx} - \frac{u}{w^2}\frac{dw}{dx}$$

This equation can be derived from earlier expressions.

Higher-order derivatives are given by

$$\frac{d^m u}{dx^m} = \frac{d^{m-1}}{dx^{m-1}}\left(\frac{du}{dx}\right)$$

The derivative $d^m u/dx^m$ is the mth order derivative of u with respect to x.

EXERCISE 6.9: Let $u = ax^2 + bx + c$. Find the second-order derivative.

Differentiation, $df(x)/dx$, is an operation d/dx on a function $f(x)$. It is possible to construct infinitesimally small differentials dx, dy by defining the differential of x, dx, as any real number in the domain $-\infty < dx < \infty$, and the differential of y, dy, as the function of x and $f(x)$ given by

$$dy = f'(x)\,dx \qquad (6.2.3)$$

If $dx = 0$, then $dy = 0$.

Suppose y is a differentiable function of x, $y = f(x)$, and x is a differentiable function of an independent variable t, $x = g(t)$. Then y is a differentiable function of t that satisfies the chain rule

$$\frac{dy}{dt} = \frac{df}{dx}\frac{dx}{dt} \qquad (6.2.4)$$

The variable t is often said to *parameterize* x and y.

EXERCISE 6.10: (a) Calculate the differential of $y = x^3$. (b) Use the chain rule to calculate dy/dx, where x, y are given by the parametric equations

$$x = 2t + 3,\, y = t^2 - 1$$

Extrema

Recall that the derivative $f'(x)$ is the slope of the curve $f(x)$ at the point x. The slope corresponding to

$$\frac{df(x)}{dx} = 0 \qquad (6.2.5)$$

can be an extremum: that is, a minimum or maximum at $x = x_{ext}$, where x_{ext} is the value of x at which Eq. (6.2.5) is valid. The type of extremum is determined by calculating the second-order derivative. If

$$\left.\frac{d^2 f(x)}{dx^2}\right|_{x_{ext}} > 0$$

then $f'(x) = 0$ corresponds to a minimum of $f(x)$ at x_{ext}. If

$$\left.\frac{d^2 f(x)}{dx^2}\right|_{x_{ext}} < 0$$

then $f'(x) = 0$ corresponds to a maximum of $f(x)$ at x_{ext}.

EXERCISE 6.11: Find the extrema of $y = ax^2 + bx + c$.

Newton–Raphson Method

Roots of the nonlinear equation $g(x) = a$, where a is a constant, can be found using the Newton–Raphson method. We first write the nonlinear equation in the form $f(x) = g(x) - a = 0$. Our problem is to find the root x_r such that $y = f(x_r) = 0$.

From Figure 6.3 we see that the slope of $f(x)$ at $x = x_0$ is

$$\text{slope} = \tan \theta = f'(x_0) = \frac{f(x_0) - 0}{(x_0 - x_1)} = \frac{f(x_0)}{(x_0 - x_1)} \tag{6.2.6}$$

The slope intersects the x axis at x_1, and $f'(x_0)$ denotes the derivative df/dx evaluated at $x = x_0$. Rearranging Eq. (6.2.6) and solving for x_1 gives

$$x_1 = x_0 - \frac{f(x_0)}{f'(x_0)}$$

The value x_1 is a first approximation of the root x_r. A better approximation can be obtained by repeating the process using x_1 as the new starting value as illustrated

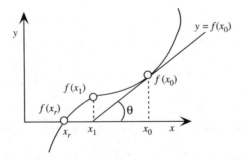

Figure 6.3   Newton–Raphson method.

in Figure 6.3. This iterative process leads to an iterative formula for converging to the answer:

$$x_{n+1} = x_n - \frac{f(x_n)}{f'(x_n)}; n = 0,1,\ldots \tag{6.2.7}$$

EXERCISE 6.12: Find the square root of any given positive number c. Show that the Newton–Raphson method works for $c = 2$.

An alternative presentation of the iterative formula is

$$f'(x_n)(x_{n+1} - x_n) = -f(x_n) \tag{6.2.8}$$

The terms $f'(x_n)$ and $f(x_n)$ are sometimes referred to as the *acceleration term* and the *residual term*, respectively. The form of Eq. (6.2.8) is suitable for extension to systems of equations.

6.3 FINITE DIFFERENCE CONCEPT

It is often relatively easy to formulate a mathematical description of a real-world problem, but not possible to solve the problem analytically. When analytical techniques are intractable, numerical methods may be the only way to make progress in finding even an approximate solution to a problem of practical importance. A frequently used numerical method is the finite difference concept described below.

Taylor Series

The basis of finite difference numerical approximations is the Taylor series. Suppose we know the value of a function $f(x)$ at $x = x_i$. We can use the Taylor series to calculate the value of $f(x)$ at $x = x_i + \Delta x$ by writing

$$f(x_i + \Delta x) = f(x_i) + \Delta x \frac{df}{dx}\bigg|_{x=x_i} + \frac{(\Delta x)^2}{2!} \frac{d^2f}{dx^2}\bigg|_{x=x_i}$$
$$+ \frac{(\Delta x)^3}{3!} \frac{d^3f}{dx^3}\bigg|_{x=x_i} + \cdots$$

where x_i and $x_i + \Delta x$ are discretized points along the x axis. The notation for a discretized representation of $f(x)$ is illustrated in Figure 6.4.

An approximation for the derivative df/dx at $x = x_i$ is obtained by rearranging the Taylor series to find

$$\frac{df}{dx} = \frac{f(x_i + \Delta x) - f(x_i)}{\Delta x} - E_T$$

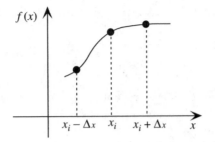

Figure 6.4 Discretizing $f(x)$.

where

$$E_T = \frac{\Delta x}{2!} \frac{d^2 f}{dx^2}\bigg|_{x=x_i} + \frac{(\Delta x)^2}{3!} \frac{d^3 f}{dx^3}\bigg|_{x=x_i} + \cdots$$

Neglecting E_T gives the finite difference approximation of the first derivative as

$$\frac{df}{dx} \approx \frac{f(x_i + \Delta x) - f(x_i)}{\Delta x}$$

The term E_T is called the *truncation error* because it contains all of the terms of the Taylor series that are neglected, or truncated, in forming an approximation of the derivative. The truncation error E_T is on the order of Δx, where the lowest order of $(\Delta x)^m$ determines the order of the finite difference approximation. The following application illustrates the use and accuracy of finite differences.

Application: Darcy's Law. The rate of fluid flow in a porous medium is given by Darcy's law [Peaceman, 1977; Aziz and Settari, 1979]; that is, flow rate is proportional to the pressure change, or

$$Q = -0.001127 \frac{KA}{\mu} \frac{dP}{dx}$$

where the symbols are defined as follows:

Q	Flow rate (bbl/day)
K	Permeability (md)
A	Cross-sectional area (ft^2)
P	Pressure (psi)
μ	Fluid viscosity (cp)
x	Length (ft)

For a porous medium, permeability is a measure of the connectedness of individual pores and is often measured in millidarcies (md). The unit of permeability is length squared. The geometry for Darcy flow is illustrated in Figure 6.5.

Suppose the physical parameters have the following values: $K = 150$ md, $\mu = 0.5$ cp, and $A = 528$ ft $\times$ 50 ft $= 2.64 \times 10^4$ ft^2. The pressure profile as a function of length is given as follows for three distances:

x (ft):	129	645	1161
P (psia):	3983	4007	4041

Notice that pressure is highest at the fluid inlet and lowest at the fluid outlet. This shows that the distance x in this example increases as we measure from left to right in Figure 6.5. Our problem is to find Q at the midpoint $x = 645$ ft.

Analytical Evaluation of dP/dx   A curve fit of the pressure profile gives the analytical expression

$$P(x) = (1.88 \times 10^{-5})x^2 + 0.03196x + 3978$$

with the derivative

$$\frac{dP}{dx} = (3.76 \times 10^{-5})x + 0.03196$$

Evaluating dP/dx at $x = 645$ ft gives the pressure gradient

$$\frac{dP}{dx} = 0.05262$$

in psia/ft. Substituting physical quantities into Darcy's law gives the flow rate

$$Q = -8926\frac{dP}{dx}$$

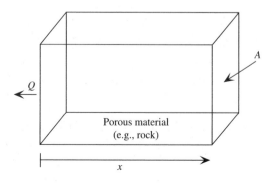

Figure 6.5   Darcy flow.

in bbl/day. At $x = 645$ ft, we have

$$Q = -502 \, \text{bbl/day}$$

Finite Difference Evaluation of dP/dx The Taylor series may be used to make the following numerical approximations:

(a) First-order correct backward difference

$$\frac{P(x) - P(x - \Delta x)}{\Delta x} = \frac{4007 - 3983}{516} = 0.0465$$

yields

$$Q = -415 \, \text{bbl/day}$$

(b) First-order correct forward difference

$$\frac{P(x + \Delta x) - P(x)}{\Delta x} = \frac{4041 - 4007}{516} = 0.0659$$

yields

$$Q = -588 \, \text{bbl/day}$$

(c) Second-order correct centered difference

$$\frac{P(x + \Delta x) - P(x - \Delta x)}{2\Delta x} = \frac{4041 - 3983}{2(516)} = 0.0562$$

yields

$$Q = -502 \, \text{bbl/day}$$

Comparing the three numerical results with the analytical value shows the expected result that the second-order correct centered difference method is more accurate than the first-order correct methods. Notice that the centered difference method requires values on both sides of the desired value; that is, from $x + \Delta x$ and $x - \Delta x$.

EXERCISE 6.13: Use the Taylor series to show that the centered difference $[P(x + \Delta x) - P(x - \Delta x)]/(2\Delta x)$ is second-order correct.

Effect of Step Size Δx If we let $x \rightarrow x + \Delta x$ in the analytical expression for $P(x)$ we find

$$\frac{\Delta P}{\Delta x} = (1.88 \times 10^{-5})(2x + \Delta x) + 0.03196$$

The effect of step size Δx is found by estimating dP/dx at $x = 645$ ft for several values of Δx:

Δx	$\Delta P/\Delta x$
516	0.0659
258	0.0611
129	0.0586
0	0.0562

As $\Delta x \to 0$ the numerical result approaches the analytical solution.

CHAPTER 7

PARTIAL DERIVATIVES

The derivative is a way to find out how a function of a single variable will change when the variable changes. We often want to extend this idea to functions of several variables. This requires the introduction of methods for calculating the effect of changing one variable while the other variables are fixed. These ideas are illustrated below.

Consider the area enclosed in Figure 7.1. The solid box encloses the original area A_0. The length of the sides of the solid box are denoted by x, y. The dotted lines in the figure represent an expansion of the original area. We wish to find the change in the total area.

Let the sides of the original box with area A_0 increase by increments Δx and Δy. First note that the original area is $A_0 = xy$ and the total area is $A_T = (x + \Delta x)(y + \Delta y)$ or $A_T = xy + x\Delta y + y\Delta x + \Delta x\Delta y$. The change in area is

$$\Delta A = A_T - A_0 = y\Delta x + x\Delta y + \Delta x\Delta y = \Delta A|_y + \Delta A|_x + \Delta x\Delta y$$

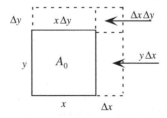

Figure 7.1 Expanding area.

Math Refresher for Scientists and Engineers, Third Edition By John R. Fanchi
Copyright © 2006 John Wiley & Sons, Inc.

where $|_y$ and $|_x$ signify that the sides of the solid box with lengths y and x are not changed. If the length y is fixed, then $\Delta A|_y = y \, \Delta x$ or $y = \Delta A / \Delta x|_y$. If the length x is fixed, $\Delta A|_x = x \, \Delta y$ or $x = \Delta A / \Delta y|_x$. Substituting these expressions into ΔA gives

$$\Delta A = \left.\frac{\Delta A}{\Delta x}\right|_y \Delta x + \left.\frac{\Delta A}{\Delta y}\right|_x \Delta y + \Delta x \, \Delta y$$

If we let $\Delta x \to 0$ and $\Delta y \to 0$, we obtain the differential of $A(x, y)$ expanded in terms of partial derivatives $(\partial A / \partial x, \partial A / \partial y)$ and differentials (dx, dy) such that

$$dA = \left.\frac{\partial A}{\partial x}\right|_y dx + \left.\frac{\partial A}{\partial y}\right|_x dy$$

where the second-order term is negligible. A more formal treatment of these ideas follows.

7.1 PARTIAL DIFFERENTIATION

A function y is a function of several variables if it is a function of two or more independent variables. A function of n variables has the form

$$y = f(x_1, x_2, \ldots, x_n)$$

where $x_1, x_2, \ldots, x_n$ are n independent variables and y is a dependent variable. The function $f(\cdot)$ is a mapping from $R^n \to R$, where R is the set of real numbers and R^n is an n-dimensional set of real numbers.

The partial derivative of y with respect to x_i is defined by

$$\frac{\partial y}{\partial x_i} = \lim_{\Delta x_i \to 0} \frac{f(x_1, \ldots, x_i + \Delta x_i, \ldots, x_n) - f(x_1, \ldots, x_i, \ldots, x_n)}{\Delta x_i}$$

and all other $\{x_j\}$ are held constant. Higher-order partial derivatives are found by successive applications of this definition; thus

$$\frac{\partial}{\partial x}\left(\frac{\partial f}{\partial x}\right) = \frac{\partial^2 f}{\partial x^2} = f_{xx}, \quad \frac{\partial}{\partial y}\left(\frac{\partial f}{\partial x}\right) = \frac{\partial^2 f}{\partial x \partial y} = f_{xy}$$

The symbol f_{xy} denotes the partial differentiation of f with respect to x and then y.

The order of differentiation is commutative. If f_{xy} and f_{yx} are continuous functions of x and y, then $f_{xy} = f_{yx}$.

Example: Let $z = ax^2 + bxy + cy^2 + dx + ey + f$.

First-Order Partial Derivatives:

$$\frac{\partial z}{\partial x} = 2ax + by + d, \quad \frac{\partial z}{\partial y} = bx + 2cy + e$$

Second-Order Partial Derivatives:

$$\frac{\partial^2 z}{\partial x^2} = \frac{\partial}{\partial x}\left(\frac{\partial z}{\partial x}\right) = 2a$$

$$\frac{\partial^2 z}{\partial y^2} = \frac{\partial}{\partial y}\left(\frac{\partial z}{\partial y}\right) = 2c$$

$$\frac{\partial^2 z}{\partial x \partial y} = \frac{\partial}{\partial x}\left(\frac{\partial z}{\partial y}\right) = \frac{\partial}{\partial y}\left(\frac{\partial z}{\partial x}\right) = b$$

EXERCISE 7.1: Let $f(x, y) = e^{xy}$. Find f_{xy}, f_{yx}.

The total differential of a function y of variables $\{x_i : i = 1, n\}$ is

$$dy = \sum_{i=1}^{n} \frac{\partial y}{\partial x_i} dx_i$$

Example: Let $y = x_1^2 + x_2^3$. In this case $n = 2$ because there are two independent variables and

$$dy = \frac{\partial y}{\partial x_1} dx_1 + \frac{\partial y}{\partial x_2} dx_2 = 2x_1 dx_1 + 3x_2^2 dx_2$$

EXERCISE 7.2: Let $y = x_1^2 + x_1^2 x_2 + x_2^3$. Calculate $\partial y/\partial x_1$, $\partial y/\partial x_2$, and dy.

Suppose we have a function z of two parameterized variables such that

$$z = f(x(t), y(t))$$

The total derivative of the parameterized function z of two variables is

$$\frac{dz}{dt} = \frac{\partial f}{\partial x}\frac{dx}{dt} + \frac{\partial f}{\partial y}\frac{dy}{dt}$$

This expression can readily be generalized for a parameterized function of several variables.

EXERCISE 7.3: Let $z = x^2 + y^2$, $x = at$, $y = be^{-t}$. Find dz/dt.

Application: Jacobian Transformation. The equation of a coordinate transformation may be written as

$$y = Ax$$

If we consider a two-dimensional system, the total derivative of the transformation equation is

$$dy_1 = \frac{\partial y_1}{\partial x_1}\, dx_1 + \frac{\partial y_1}{\partial x_2}\, dx_2$$

$$dy_2 = \frac{\partial y_2}{\partial x_1}\, dx_1 + \frac{\partial y_2}{\partial x_2}\, dx_2$$

or

$$dy_1 = J_{11} dx_1 + J_{12} dx_2$$
$$dy_2 = J_{21} dx_1 + J_{22} dx_2$$

where

$$J_{ij} = \frac{\partial y_i}{\partial x_j}$$

Collecting the elements $\{J_{ij}\}$ in the 2×2 square matrix J gives

$$[dy]_i = \sum_{j=1}^{2} [J]_{ij}\, [dx]_j$$

or the matrix equation

$$dy = J dx$$

The matrix J is called the Jacobian and is given by

$$[J]_{if} = \frac{\partial y_i}{\partial x_j}$$

EXERCISE 7.4: Find the Jacobian of the coordinate rotation $y = ax$, where a is the matrix of coordinate rotations introduced in Chapter 4, Section 4.1.

7.2 VECTOR ANALYSIS

Vector analysis is the study of vectors and their transformation properties. As we show in the following, it is a discipline in which partial differentiation plays a major role.

Scalar and Vector Fields

Let x_1, x_2, x_3 denote the Cartesian coordinates of a point X in a region of space R. The position vector x of X is

$$x = x_1 i_1 + x_2 i_2 + x_3 i_3$$

where i_1, i_2, i_3 are unit vectors defined along the orthogonal axes of the coordinate system. If we can associate a scalar function f with every point in R, then f is the scalar field in R and may be written

$$f(x_1, x_2, x_3) = f(x)$$

An example of a scalar field is the temperature at each point in a region of space.

Instead of a scalar function, suppose we associate a vector v with every point in R. The resulting vector field has the form

$$v(x_1, x_2, x_3) = v(x)$$

A vector field is exemplified by a velocity field or a magnetic field. The vector field is a function that assigns a vector to every point in a region.

Scalar and vector fields commute, that is,

$$fu = uf$$

Vector fields u and v may be multiplied in the usual way as the dot product

$$u \cdot v = (u_1 i_1 + u_2 i_2 + u_3 i_3) \cdot (v_1 i_1 + v_2 i_2 + v_3 i_3) = \sum_{m=1}^{3} u_m v_m$$

and the cross product

$$u \times v = \begin{vmatrix} i_1 & i_2 & i_3 \\ u_1 & u_2 & u_3 \\ v_1 & v_2 & v_3 \end{vmatrix} = -v \times u$$

Vector fields u, v, w satisfy the triple scalar product

$$u \cdot (v \times w) = v \cdot (w \times u) = w \cdot (u \times v)$$

and the triple vector product

$$u \times (v \times w) = v(u \cdot w) - (u \cdot v)$$

Example: The expansion of the triple scalar product of the vector fields A, B, C gives

$$A \cdot (B \times C) = (a_x\hat{i} + a_y\hat{j} + a_z\hat{k}) \cdot \begin{vmatrix} \hat{i} & \hat{j} & \hat{k} \\ b_x & b_y & b_z \\ c_x & c_y & c_z \end{vmatrix}$$

$$= (a_x\hat{i} + a_y\hat{j} + a_z\hat{k}) \cdot [(b_yc_z - b_zc_y)\hat{i} - (b_xc_z - b_zc_x)\hat{j}$$
$$+ (b_xc_y - b_yc_x)\hat{k}]$$
$$= a_x(b_yc_z - b_zc_y) + a_y(b_zc_x - b_xc_z) + a_z(b_xc_y - b_yc_x)$$

Gradient, Divergence, and Curl

We can determine the spatial variation of a scalar or vector field by introducing the *del operator* ∇ defined in Cartesian coordinates as

$$\nabla \equiv i_1 \frac{\partial}{\partial x_1} + i_2 \frac{\partial}{\partial x_2} + i_3 \frac{\partial}{\partial x_3}$$

Applying ∇ to the scalar field f gives a vector field called the *gradient* of f. It is denoted as

$$\text{grad } f = \nabla f = i_1 \frac{\partial f}{\partial x_1} + i_2 \frac{\partial f}{\partial x_2} + i_3 \frac{\partial f}{\partial x_3}$$

The gradient of f points in the direction in which the scalar field f changes the most with respect to a change in position. The vector field ∇f is perpendicular, or normal, to the surfaces corresponding to constant values of f. The arrows in Figure 7.2 illustrate the direction of the gradient at several points around the surface of constant f.

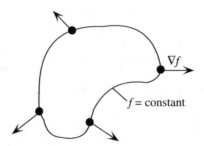

Figure 7.2 Direction of gradient.

Example: Let $f(x, y) = x^2 y$ be a scalar field. The gradient of $f(x, y)$ is

$$\nabla f(x, y) = \hat{i}\frac{\partial f}{\partial x} + \hat{j}\frac{\partial f}{\partial y} = 2xy\hat{i} + x^2\hat{j}$$

The del operator can be applied to vector fields in two ways: (1) to create a scalar and (2) to create another vector. Suppose a is a vector field. We create a scalar by defining the divergence of a as the dot product of the operator ∇ and the vector a:

$$\text{div } a \equiv \nabla \cdot a = \left(i_1\frac{\partial}{\partial x_1} + i_2\frac{\partial}{\partial x_2} + i_3\frac{\partial}{\partial x_3}\right) \cdot (a_1 i_1 + a_2 i_2 + a_3 i_3)$$

$$= \frac{\partial a_1}{\partial x_1} + \frac{\partial a_2}{\partial x_2} + \frac{\partial a_3}{\partial x_3}$$

because $i_m \cdot i_n = \delta_{mn}$, where δ_{mn} is the Kronecker delta. The cross product of the operator ∇ and the vector a is called the *rotation* of a or the *curl* of a. It is given by

$$\text{curl } a \equiv \nabla \times a = \begin{vmatrix} i_1 & i_2 & i_3 \\ \dfrac{\partial}{\partial x_1} & \dfrac{\partial}{\partial x_2} & \dfrac{\partial}{\partial x_3} \\ a_1 & a_2 & a_3 \end{vmatrix}$$

Note that $a \cdot \nabla$ and $a \times \nabla$ are not commutative; that is, $a \cdot \nabla \neq \nabla \cdot a$ and $a \times \nabla \neq \nabla \times a$. The vector products $a \cdot \nabla$ and $a \times \nabla$ are given by

$$a \cdot \nabla = (a_1 i_1 + a_2 i_2 + a_3 i_3) \cdot \left(i_1\frac{\partial}{\partial x_1} + i_2\frac{\partial}{\partial x_2} + i_3\frac{\partial}{\partial x_3}\right)$$

$$= \sum_{m=1}^{3} a_m \frac{\partial}{\partial x_m}$$

and

$$a \times \nabla = \begin{vmatrix} i_1 & i_2 & i_3 \\ a_1 & a_2 & a_3 \\ \dfrac{\partial}{\partial x_1} & \dfrac{\partial}{\partial x_2} & \dfrac{\partial}{\partial x_3} \end{vmatrix}$$

The divergence of the gradient of a scalar field f gives the Laplacian of f; thus in Cartesian coordinates we have

$$\nabla \cdot (\nabla f) = \sum_{m=1}^{3} \frac{\partial^2 f}{\partial x_m^2} \equiv \nabla^2 f$$

By contrast, the Laplacian of a vector field u is

$$\nabla^2 u = \nabla(\nabla \cdot u) - \nabla \times (\nabla \times u)$$

The curl of the gradient of a scalar field vanishes,

$$\nabla \times (\nabla f) = \mathbf{0}$$

and the divergence of the curl of a vector field is zero,

$$\nabla \cdot (\nabla \times u) = 0$$

Several other useful relations are summarized as follows:

DEL OPERATOR RELATIONS

Let f, g be scalar fields and u, v be vector fields.

Sum of fields	$\nabla(f + g) = \nabla f + \nabla g$
	$\nabla \cdot (u + v) = \nabla \cdot u + \nabla \cdot v$
	$\nabla \times (u + v) = \nabla \times u + \nabla \times v$
Product of fields	$\nabla(fg) = f(\nabla g) + g(\nabla f)$
	$\nabla \cdot (fu) = f(\nabla \cdot u) + (\nabla f) \cdot u$
	$\nabla \times (fu) = f(\nabla \times u) + (\nabla f) \times u$
	$\nabla \cdot (u \times v) = v \cdot (\nabla \times u) - u \cdot (\nabla \times v)$
	$\nabla \times (u \times v) = u(\nabla \cdot v) + (v \cdot \nabla)$
	$\quad u - v(\nabla \cdot u) - (u \cdot \nabla)v$
	$\nabla(u \cdot v) = u \times (\nabla \times v) - v(\nabla \cdot u) +$
	$\quad (v \cdot \nabla) u - (u \cdot \nabla)v$
Laplacian	$\nabla \cdot (\nabla f) = \nabla^2 f$
	$\nabla \times (\nabla \times u) = \nabla(\nabla \cdot u) - \nabla^2 u$

Example: Suppose $A = 2xy\hat{i} + x^2\hat{j} = A_x\hat{i} + A_y\hat{j}$. Then the dot product is

$$\nabla \cdot A = \frac{\partial}{\partial x}A_x + \frac{\partial}{\partial y}A_y = 2y$$

The cross product is

$$\nabla \times A = \begin{vmatrix} \hat{i} & \hat{j} & \hat{k} \\ \dfrac{\partial}{\partial x} & \dfrac{\partial}{\partial y} & \dfrac{\partial}{\partial z} \\ A_x & A_y & A_z \end{vmatrix}$$

Expanding the determinant and simplifying gives

$$\nabla \times A = \hat{i}\left(\frac{\partial A_z}{\partial y} - \frac{\partial A_y}{\partial z}\right) + \hat{j}\left(\frac{\partial A_x}{\partial z} - \frac{\partial A_z}{\partial x}\right) + \hat{k}\left(\frac{\partial A_y}{\partial x} - \frac{\partial A_x}{\partial y}\right) = -2x\hat{k}$$

The curl of A is a vector transverse to the $x-y$ plane containing the vector A. Also, the vector A is the gradient of $f(x, y) = x^2y$ from a previous example; thus

$$\nabla \cdot A = \nabla \cdot (\nabla f) = \nabla^2 f = \frac{\partial^2 f}{\partial x^2} + \frac{\partial^2 f}{\partial y^2} + \frac{\partial^2 f}{\partial z^2} = 2y$$

A field vector v is called *irrotational* if the curl, or rotation, of v vanishes, that is,

$$\nabla \times v = 0$$

The vector field v is said to be *solenoidal* if

$$\nabla \cdot v = 0$$

A vector field V that is the gradient of a scalar field ϕ is irrotational because the curl of a gradient vanishes; thus

$$\nabla \times V = \nabla \times (\nabla \phi) = \mathbf{0}$$

Similarly, a vector field U that is the curl of a vector field u is solenoidal because the divergence of the curl vanishes; thus

$$\nabla \cdot U = \nabla \cdot (\nabla \times u) = 0$$

EXERCISE 7.5: Let $r = x\hat{i} + y\hat{j} + z\hat{k}$. Evaluate (a) $r = |r|$; (b) $\nabla \cdot r$; (c) $\nabla \times r$; (d) $\nabla(r)$; and (e) $\nabla(1/r)$.

Application: Propagation of Seismic Waves. Seismic waves are vibrations, or displacements from an undisturbed position, that propagate from a source, such as an explosion, through the earth. Seismic wave propagation is an example of a displacement propagating through an elastic medium. The equation for a wave propagating through an elastic, homogeneous, isotropic medium is

$$\rho \frac{\partial^2 u}{\partial t^2} = (\lambda + 2\mu) \nabla (\nabla \cdot u) - \mu \nabla \times (\nabla \times u) \tag{7.2.1}$$

where ρ is the mass density of the medium, λ and μ are properties of the elastic medium called *Lamé's constants*, and u measures the displacement of the medium from its undisturbed state [Tatham and McCormack, 1991].

If the displacement u_I is irrotational, then u_I satisfies the constraint

$$\nabla \times u_I = 0$$

and Eq. (7.2.1) becomes

$$\frac{\partial^2 \boldsymbol{u}_I}{\partial t^2} = \frac{(\lambda + 2\mu)}{\rho} \nabla(\nabla \cdot \boldsymbol{u}_I) \tag{7.2.2}$$

The vector identity

$$\nabla(\nabla \cdot \boldsymbol{u}) = \nabla \times (\nabla \times \boldsymbol{u}) + \nabla^2 \boldsymbol{u} \tag{7.2.3}$$

for an irrotational vector is

$$\nabla(\nabla \cdot \boldsymbol{u}_I) = \nabla^2 \boldsymbol{u}_I$$

so that Eq. (7.2.1) becomes the wave equation

$$\frac{\partial^2 \boldsymbol{u}_I}{\partial t^2} = v_I^2 \, \nabla^2 \boldsymbol{u}_I \tag{7.2.4}$$

The speed of wave propagation v_I for an irrotational displacement $\boldsymbol{u}_I$ is

$$v_I = \sqrt{\frac{(\lambda + 2\mu)}{\rho}}$$

A solution of Eq. (7.2.4) that is irrotational is the solution for a longitudinal wave propagating in the z direction with amplitude u_0, frequency ω, and wavenumber k:

$$\boldsymbol{u}_I = u_0 e^{i(kz - \omega t)} \hat{k}$$

If the displacement $\boldsymbol{u}_s$ is solenoidal, then $\boldsymbol{u}_s$ satisfies the constraint

$$\nabla \cdot \boldsymbol{u}_s = 0$$

and Eq. (7.2.1) becomes

$$\frac{\partial^2 \boldsymbol{u}_s}{\partial t^2} = -\frac{\mu}{\rho} \, \nabla \times (\nabla \times \boldsymbol{u}_s) \tag{7.2.5}$$

The vector identity in Eq. (7.2.3) for a solenoidal vector reduces to

$$\nabla \times (\nabla \times \boldsymbol{u}_s) = -\nabla^2 \boldsymbol{u}_s$$

so that Eq. (7.2.5) becomes the wave equation

$$\frac{\partial^2 \boldsymbol{u}_s}{\partial t^2} = v_s^2 \, \nabla^2 \boldsymbol{u}_s \tag{7.2.6}$$

The speed of wave propagation v_s for a solenoidal displacement $\boldsymbol{u}_s$ is

$$v_s = \sqrt{\frac{\mu}{\rho}} < v_I$$

A solution of Eq. (7.2.6) that is solenoidal is the solution for a transverse wave propagating in the z direction:

$$\boldsymbol{u}_s = u_0 e^{i(kx - \omega t)} \hat{k}$$

The irrotational displacement $\boldsymbol{u}_I$ represents a longitudinal P (primary) wave, whereas the solenoidal displacement $\boldsymbol{u}_s$ represents a slower transverse S (secondary) wave. Both types of waves are associated with earthquakes and explosions on or below the earth's surface. The waves are useful for geophysical studies of the earth's interior.

7.3 ANALYTICITY AND THE CAUCHY–RIEMANN EQUATIONS

A fundamental concept in complex analysis is the concept of analyticity. A function $f(z)$ of a complex variable z is analytic in a domain D if $f(z)$ exists and is differentiable at all points in D. In principle, the complex function $f(z)$ of the complex variable $z = x + iy$ can be written as the sum of a real function $u(x, y)$ and an imaginary function $iv(x, y)$; thus

$$f(z) = u(x, y) + iv(x, y) \qquad (7.3.1)$$

where u and v are real functions of the real variables x, y.

The criteria for establishing the analyticity of $f(z)$ are obtained by extending the definition of partial derivative presented in Section 7.1 from real to complex functions. The derivative of $f(z)$ with respect to z is

$$f'(z) = \lim_{\Delta z \to 0} \frac{f(z + \Delta z) - f(z)}{\Delta z} \qquad (7.3.2)$$

The complex increment $\Delta z = \Delta x + i\Delta y$ may approach 0 along any path in a neighborhood of z. A neighborhood of z is an open set in the complex plane that encloses the point z. One example path for $\Delta z \to 0$ is to approach the point z by letting $\Delta x \to 0$ and then $\Delta y \to 0$. In this case, $\Delta z \to i\Delta y$ and Eq. (7.3.2) becomes

$$f'(z) = \lim_{\Delta y \to 0} \frac{f(x, y + \Delta y) - f(x, y)}{i\Delta y} \qquad (7.3.3)$$

Expanding Eq. (7.3.3) in terms of u, v gives

$$f'(z) = \lim_{\Delta y \to 0} \frac{u(x, y + \Delta y) + iv(x, y + \Delta y) - u(x, y) - iv(x, y)}{i\,\Delta y} \qquad (7.3.4)$$

or

$$f'(z) = \lim_{\Delta y \to 0} \frac{v(x, y + \Delta y) - v(x, y)}{\Delta y}$$
$$- i \lim_{\Delta y \to 0} \frac{u(x, y + \Delta y) - u(x, y)}{\Delta y}$$

(7.3.5)

Applying the definition of partial derivative in Section 7.1 to Eq. (7.3.5) yields

$$f'(z) = \frac{\partial v}{\partial y} - i \frac{\partial u}{\partial y}$$

(7.3.6)

An alternative route to an expression for $f'(z)$ is provided by letting $\Delta y \to 0$ before letting $\Delta x \to 0$. In this case, $\Delta z \to \Delta x$ and Eq. (7.3.2) becomes

$$f'(z) = \lim_{\Delta x \to 0} \frac{f(x + \Delta x, y) - f(x, y)}{\Delta x}$$

(7.3.7)

By analogy with the procedure leading from Eq. (7.3.3) to (7.3.6), we obtain

$$f'(z) = \frac{\partial u}{\partial x} + i \frac{\partial v}{\partial x}$$

(7.3.8)

Equations (7.3.6) and (7.3.8) are equivalent expressions for $f'(z)$.

Equating the real and imaginary parts of Eqs. (7.3.6) and (7.3.8) gives the Cauchy–Riemann equations

$$\frac{\partial u}{\partial x} = \frac{\partial v}{\partial y}$$

(7.3.9)

and

$$-\frac{\partial u}{\partial y} = \frac{\partial v}{\partial x}$$

(7.3.10)

The validity of the Cauchy–Riemann equations depends on the existence of $f'(z)$, which depends on the analyticity of $f(z)$. If the function $f(z)$ is analytic, then $f'(z)$ exists and the Cauchy–Riemann equations are valid. Conversely, if the Cauchy–Riemann equations apply, then the function $f(z)$ is analytic and $f'(z)$ exists [Kreyszig, 1999].

If we differentiate Eq. (7.3.9) with respect to x and Eq. (7.3.10) with respect to y, we obtain

$$\frac{\partial^2 u}{\partial x^2} = \frac{\partial^2 v}{\partial x \partial y}$$

(7.3.11)

and

$$-\frac{\partial^2 u}{\partial y^2} = \frac{\partial^2 v}{\partial y \partial x} \tag{7.3.12}$$

Subtracting Eq. (7.3.12) from (7.3.11) gives Laplace's equation in two Cartesian dimensions:

$$\frac{\partial^2 u}{\partial x^2} + \frac{\partial^2 u}{\partial y^2} = 0 \tag{7.3.13}$$

Laplace's equation is an example of a partial differential equation. Partial differential equations are discussed in more detail in Chapter 12.

EXERCISE 7.6: Given $f(z) = z^2$, show that the Cauchy–Riemann equations are satisfied and evaluate $f'(z)$.

CHAPTER 8

INTEGRAL CALCULUS

Differential calculus may be thought of as a reductionist view of change: we study a changing system by looking at infinitesimally small variations. The inverse process is to reconstruct the whole by summing the infinitesimal differences using the apparatus of integral calculus.

8.1 INDEFINITE INTEGRALS

The indefinite integral is the antiderivative, or inverse, of the derivative. Suppose the function $g(x)$ is given and has the derivative

$$g'(x) = \frac{dg(x)}{dx}$$

Taking the inverse of $g'(x)$ should recover $g(x)$; thus

$$[g'(x)]^{-1} = \int [g'(x)]dz = g(x) + C$$

Notice that an additive constant C has appeared on the right-hand side of the equation. This constant appears because the derivative of $g(x) + C$ is the same as the derivative of $g(x)$, so information may be lost upon differentiation. The constant

Math Refresher for Scientists and Engineers, Third Edition By John R. Fanchi
Copyright © 2006 John Wiley & Sons, Inc.

C must be determined from additional information, such as specifying $g(x)$ at a particular value of x.

Example: Let $g(x) = x^n$ so that

$$g'(x) = nx^{n-1}$$

The integral of $g'(x)$ is

$$\int g'(x)dx = \int nx^{n-1}dx = x^n + C$$

The integration constant C must equal 0 in this case since $g(x) = x^n$.

For two continuous functions $f(x)$, $g(x)$ we have the properties

$$\int [f(x) \pm g(x)]dx = \int f(x)dx \pm \int g(x)dx$$

and

$$\int af(x)dx = a\int f(x)dx$$

where a is a constant. The variable of integration is a dummy variable, that is,

$$\int f(x)dx = \int f(u)du$$

where x, u are dummy variables that may be freely interchanged. From the definition of the indefinite integral as the inverse of the derivative, we find that the derivative of an indefinite integral is the integrand; thus

$$\frac{d}{dx}\left[\int f(x)dx\right] = f(x)$$

Example: Suppose $g(x) = x + b$, where b is a constant. Then $g'(x) = 1$. The indefinite integral of $g'(x)$ is

$$g(x) = \int g'(x)dx = \int 1dx = x + C$$

where C is the integration constant. Comparing the integrated function with the original function shows that $C = b$. Alternatively, we can say that C is determined by specifying the condition $g(0) = b$.

Tables of Integrals

Integrals of well-known functions have been tabulated in handbooks. These integrals were evaluated using several different strategies. One useful strategy is to calculate the derivative of a function, then recognize that the inverse operation of differentiation is integration. A few of these integrals are given below.

INTEGRALS

Assume (u, v) are functions of x and (a, n) are constants. An integration constant C should be added to each indefinite integral.

$$\int a\,dx = ax$$

$$\int u\,dv = uv - \int v\,du$$

$$\int ax^n\,dx = a\frac{x^{n+1}}{n+1}, n \neq -1$$

$$\int \frac{a}{x}\,dx = a\ln|x|$$

$$\int \sin ax\,dx = -\frac{1}{a}\cos ax$$

$$\int \sinh ax\,dx = \frac{1}{a}\cosh ax$$

$$\int \cos ax\,dx = \frac{1}{a}\sin ax$$

$$\int \cosh ax\,dx = \frac{1}{a}\sinh ax$$

$$\int \tan ax\,dx = -\frac{1}{a}\ln|\cos ax|$$

$$\int \tanh ax\,dx = \frac{1}{a}\ln(\cosh ax)$$

$$\int e^{ax}\,dx = \frac{1}{a}e^{ax}$$

$$\int \ln ax\,dx = x\ln ax - x$$

$$\int x\sin ax\,dx = \frac{1}{a^2}\sin ax - \frac{x}{a}\cos ax$$

$$\int x\cos ax\,dx = \frac{1}{a^2}\cos ax + \frac{x}{a}\sin ax$$

$$\int \sin^2 ax\,dx = \frac{x}{2} - \frac{\sin 2ax}{4a}$$

$$\int \cos^2 ax\,dx = \frac{x}{2} + \frac{\sin 2ax}{4a}$$

$$\int \sinh^2 ax\,dx = \frac{\sinh 2ax}{4a} - \frac{x}{2}$$

$$\int \cosh^2 ax\,dx = \frac{\sinh 2ax}{4a} + \frac{x}{2}$$

$$\int x^n e^{ax}\,dx = \frac{1}{a}x^n e^{ax} - \frac{n}{a}\int x^{n-1}e^{ax}\,dx$$

$$\int x^n \ln ax\,dx = \frac{x^{n+1}}{n+1}\ln ax - \frac{x^{n+1}}{(n+1)^2}$$

$$\int \frac{dx}{a^2+x^2} = \frac{1}{a}\tan^{-1}\frac{x}{a}$$

$$\int \frac{dx}{a^2-x^2} = \frac{1}{2a}\ln\left|\frac{a+x}{a-x}\right|$$

8.2   DEFINITE INTEGRALS

Let $f(x)$ be continuous and single valued for $a \leq x \leq b$. Divide the interval (a, b) of the x axis into n parts at the points $(a \equiv x_0), x_1, x_2, \ldots, (x_n \equiv b)$ as in Figure 8.1. Define $\Delta x_i = x_i - x_{i-1}$ and let ξ_i be a value of x in the interval $x_{i-1} < \xi_i \leq x_i$. Form the sum of the areas of each approximately rectangular area to get

$$\sum_{i=1}^{n} f(\xi_i)\Delta x_i$$

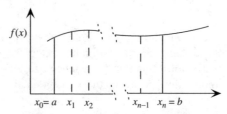

Figure 8.1 Discretizing the area under the curve.

If we take the limit as $n \to \infty$ and $\Delta x_i \to 0$, the approximation of the area under the curve as a rectangle becomes more accurate and we obtain the definition of the definite integral of $f(x)$ between the limits a, b; thus

$$\lim_{\substack{n \to \infty \\ \Delta x_i \to 0}} \sum_{i=1}^{n} f(\xi_i) \Delta x_i \equiv \int_a^b f(x)dx \tag{8.2.1}$$

The points a, b are referred to as the *lower* and *upper limits* of integration, respectively. The definite integral

$$\int_a^b f(x)\,dx$$

equals the area under the curve $f(x)$ in the interval (a, b).

Fundamental Theorem of Calculus

The connection between differential and integral calculus is through the fundamental theorem of calculus [Larson et al., 1990]. In particular, if $f(x)$ is continuous in the interval $a \le x \le b$ and $G(x)$ is a function such that $dG/dx = f(x)$ for all values of x in (a, b), then

$$\int_a^b f(x)dx = G(b) - G(a) \tag{8.2.2}$$

The definite integral acts like an antiderivative of $f(x)$, or the inverse operation to differentiation. The concept of inverse differentiation can be made more explicit by introducing the notation

$$[G(x)]_a^b = G(b) - G(a) \tag{8.2.3}$$

and using $dG/dx = f(x)$ in the definite integral of Eq. (8.2.2) to give

$$\int_a^b \left(\frac{dG}{dx}\right) dx = [G(x)]_a^b \tag{8.2.4}$$

Relation of Definite and Indefinite Integrals

We can express the indefinite integral as a definite integral with a variable limit:

$$\int f(x)\,dx = \int_a^x f(u)\,du + C'$$

where C' is a constant of integration.

Example: Let $f(x) = \sin x$. The indefinite integral of the left-hand side is

$$\int f(x)\,dx = -\cos x + C$$

Consider the right-hand side:

$$\int_a^x f(u)\,du + C' = -[\cos u]_a^x + C' = -\cos x + \cos a + C'$$

Comparing the left- and right-hand sides implies $C = \cos a + C'$.

Derivative of Definite Integrals

Let $\phi(\alpha)$ be the definite integral

$$\phi(\alpha) = \int_{u(\alpha)}^{v(\alpha)} f(y, \alpha)\,dy \qquad (8.2.5)$$

where $f(y, \alpha)$ is a function of the independent variables y and α. The limits $u(\alpha)$, $v(\alpha)$ are differentiable functions of the independent variable α, and α is defined in the interval $a \le \alpha \le b$, where a and b are constants. Furthermore, suppose $f(y, \alpha)$ and $\partial f / \partial \alpha$ are continuous in the $y\alpha$ plane defined by the intervals $u \le y \le v$ and $a \le \alpha \le b$. Under these conditions, the derivative of $\phi(\alpha)$ with respect to α is given by Leibnitz's rule [Arfken and Weber, 2001; Spiegel, 1963]:

$$\frac{d}{d\alpha}\phi(\alpha) = \int_{u(\alpha)}^{v(\alpha)} \frac{\partial f(y,\alpha)}{\partial \alpha}\,dy$$
$$+ f(v(\alpha), \alpha)\frac{dv(\alpha)}{d\alpha} - f(u(\alpha), \alpha)\frac{du(\alpha)}{d\alpha} \qquad (8.2.6)$$

If u and v are equated to constants c and d, respectively, Eq. (8.2.6) becomes

$$\frac{d}{d\alpha}\int_c^d f(y, \alpha)\,dy = \int_c^d \frac{\partial f(y,\alpha)}{\partial \alpha}\,dy \qquad (8.2.7)$$

If the function f does not depend on α, Eq. (8.2.6) becomes

$$\frac{d}{d\alpha}\int_{u(\alpha)}^{v(\alpha)} f(y)dy = f(v(\alpha))\frac{dv(\alpha)}{d\alpha} - f(u(\alpha))\frac{du(\alpha)}{d\alpha} \tag{8.2.8}$$

Example: Let $f(y) = \ln y$, $u(\alpha) = a$, and $v(\alpha) = \alpha$. In this case, the function f does not depend on α. Consequently, we substitute u, v, and f into Eq. (8.2.8) to find

$$\frac{d}{d\alpha}\int_a^\alpha \ln y \, dy = f(\alpha)\frac{d\alpha}{d\alpha} - f(a)\frac{da}{d\alpha} = \ln \alpha \tag{8.2.9}$$

because $da/d\alpha = 0$ and $f(\alpha) = \ln\alpha$. This result is verified by first evaluating the integral

$$\int_a^\alpha \ln y \, dy = [y\ln y - y]_a^\alpha = \alpha\ln\alpha - \alpha - (a\ln a - a) \tag{8.2.10}$$

Taking the derivative of Eq. (8.2.10) with respect to α gives

$$\frac{d}{d\alpha}[\alpha\ln\alpha - \alpha - (a\ln a - a)] = \frac{d}{d\alpha}[\alpha\ln\alpha - \alpha]$$
$$= \ln\alpha \tag{8.2.11}$$

Equation (8.2.11) agrees with Eq. (8.2.9) and verifies Eq. (8.2.8).

8.3 SOLVING INTEGRALS

Change of Variable or Variable Substitution

One of the most powerful methods for solving integrals is to make a change of variables until the original integral acquires the form of an integral with a known solution. This technique is also known as variable substitution because one variable is being substituted for another. To substitute one variable for another, let $x = g(t)$. We can change the integral of a function with respect to x into an integral with respect to the new variable t by making a change of variables. For an indefinite integral we obtain

$$\int f(x)\,dx = \int f(g(t))\left[\frac{dg(t)}{dt}\right]dt$$

The definite integral is a little more complicated because we have to account for a change of integral limits. The result is

$$\int_a^b f(x)\,dx = \int_\alpha^\beta f(g(t))\left[\frac{dg(t)}{dt}\right]dt$$

where the integration limits are $g(\alpha) = a$ and $g(\beta) = b$.

EXERCISE 8.1: Given

$$f(x) = \frac{1}{x}, x = g(t) = t^n$$

Find $\int_a^b f(x)\,dx$ by direct integration and by a change of variable.

Integration by Parts

Consider the differential of a product of functions u, v:

$$d(uv) = u\,dv + v\,du \qquad (8.3.1)$$

Integrate Eq. (8.3.1) between the integration limits A and B to get

$$\int_A^B d(uv) = \int_A^B u\,dv + \int_A^B v\,du \qquad (8.3.2)$$

Evaluating the left-hand side gives

$$[uv]_A^B = \int_A^B u\,dv + \int_A^B v\,du \qquad (8.3.3)$$

Rearranging Eq. (8.3.3) gives the formula for an integration by parts, namely,

$$\int_A^B u\,dv = [uv]_A^B - \int_A^B v\,du \qquad (8.3.4)$$

EXERCISE 8.2: Evaluate the indefinite integral $I(x) = \int x\sin x\,dx$ using integration by parts.

Application: Length of a Curve. Consider the curve in Figure 8.2 between the end points A and B. We approximate the curve with straight-line segments. The length of the curve is approximately the sum of the straight-line segments;

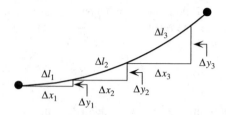

Figure 8.2 Length of a curve.

that is, $L \approx \Sigma_i \Delta l_i$. From the Pythagorean theorem, we find the length of a line segment is

$$\Delta l_i = \sqrt{\Delta x_i^2 + \Delta y_i^2} = \left[1 + \left(\frac{\Delta y_i}{\Delta x_i} \right)^2 \right]^{1/2} \Delta x_i$$

The curve length L is given by the sum over the length of all line segments:

$$L \approx \sum_i \left[1 + \left(\frac{\Delta y_i}{\Delta x_i} \right)^2 \right]^{1/2} \Delta x_i$$

In the limit as $\Delta x_i \to 0$ we have

$$L = \int_A^B \left[1 + \left(\frac{dy}{dx} \right)^2 \right]^{1/2} dx$$

where A, B are (boundary) end points of the curve.

EXERCISE 8.3: Calculate the length of a straight line $y = mx + b$ between the end points x_A, x_B.

8.4 NUMERICAL INTEGRATION

Integrals occur in many applications, yet it is not always possible to solve them analytically. When the integral is intractable, it is necessary to use an approximation technique.

Definite Integrals

Several numerical techniques exist for solving definite integrals. One widely used technique is to approximate the definite integral with an nth order polynomial. The definite integral is then approximated by the sum

$$\int_a^b f(x) \, dx \approx \sum_{i=0}^n w_i f(x_i) \tag{8.4.1}$$

where the finite interval $[a, b]$ is divided into n intervals bounded by $n + 1$ evenly spaced base points $\{x_0, x_1, \ldots, x_n\}$ with separation

$$h = x_{i+1} - x_i = \frac{b - a}{n}$$

A weight factor w_i is associated with the functional value $f(x_i)$.

Trapezoidal Rule

The trapezoidal rule approximates the integral $f(x)$ of Eq. (8.4.1) with a straight line between the limits x_i, x_{i+1} of a subinterval. Since it takes two points on a graph to define a straight line, the trapezoidal rule is sometimes called the *2-point* rule. The trapezoidal rule has the form

$$\int_a^b f(x)\, dx \approx h\left[\tfrac{1}{2}f(a) + f(x_1) + f(x_2) + \cdots + f(x_{n-1}) + \tfrac{1}{2}f(b) \right] \qquad (8.4.2)$$

where $x_0 = a$, $x_n = b$, $\{x_{i+1} = x_i + h$ for all $i = 0, 1, \ldots, n - 1\}$, and $h = (b - a)/n$.

Example: The simplest application of the trapezoidal rule is the case in which only one interval is defined, that is, $n = 1$. Then Eq. (8.4.2) becomes

$$\int_a^b f(x)\, dx \approx h\left[\tfrac{1}{2}f(a) + \tfrac{1}{2}f(b) \right]$$

Simpson's Rule

A more accurate numerical integration algorithm than the trapezoidal rule is obtained by approximating the integrand $f(x)$ of Eq. (8.4.1) with a second-order polynomial (a quadratic equation) between the limits x_i, x_{i+2} of a pair of subintervals. The equally spaced subintervals are paired beginning with $i = 0$. To assure that every subinterval has a pair, the number of subintervals n must be even. Each double subinterval then contains three points on a graph. For this reason, Simpson's rule is sometimes called the *3-point* rule.

Simpson's rule has the form

$$\int_a^b f(x)\, dx \approx \frac{h}{3}[f(x_0) + 4f(x_1) + 2f(x_2)$$

$$+ 4f(x_3) + \cdots + 4f(x_{n-1}) + f(x_n)] \qquad (8.4.3)$$

where $x_0 = a$, $x_n = b$, $\{x_{i+1} = x_i + h$ for all $i = 0, 1, \ldots, n - 1\}$, and $h = (b - a)/n$.

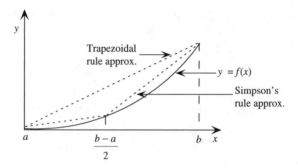

Figure 8.3 Numerical integration.

Example: The simplest application of Simpson's rule is the case in which two subintervals are defined in the interval between a and b. Thus $n = 2$ and Eq. (8.4.3) becomes

$$\int_a^b f(x)\,dx \approx \frac{h}{3}[f(a) + 4f(a+h) + f(b)]$$

where $h = (b-a)/2$. Notice that the number of subintervals $n = 2$ is even.

The trapezoidal rule approximates the integrand with a straight line between the limits a, b. Simpson's rule uses a polynomial of order 2, a quadratic function, to approximate the integrand. Simpson's rule requires more work than the trapezoidal rule but is more accurate. This is illustrated in Figure 8.3 and Exercise 8.4.

EXERCISE 8.4: Let $y = f(x) = x^2$. Evaluate the definite integral $\int_0^b f(x)\,dx$ analytically and numerically using the trapezoidal rule and Simpson's rule. Compare the results.

Improper Integrals

Methods exist for numerically integrating improper integrals in which the magnitude of one or both limits is ∞. The quadrature formulas are obtained by fitting the integrand to polynomials with unequal spacing between base points $\{x_i\}$. These methods are presented in references such as Carnahan et al. [1969] and Press et al. [1992].

CHAPTER 9

SPECIAL INTEGRALS

The fundamental concepts of integral calculus have been applied in many areas of mathematics. Some of these applications are considered here as special integrals. A few strategies for evaluating special integrals are illustrated below for the line integral and the double integral. These integrals provide an introduction to vector calculus integrals and multiple integrals, respectively. We then present three transforms: the Fourier transform, Z transform, and Laplace transform. The Fourier transform and Laplace transform are integral transforms, and the Z transform is related to the Fourier transform.

9.1 LINE INTEGRAL

The line integral may be viewed as an extension of the concept of definite integral

$$I_d = \int_a^b f(x)dx$$

In the case of I_d, we integrate $f(x)$ along the x axis between the points $x = a$ and $x = b$. Suppose that instead of integrating along the x axis, we integrate along a piecewise smooth curve C. The curve C is piecewise smooth in the interval $[a, b]$ if the interval can be partitioned into a finite number of subintervals on which C is smooth, that is, continuously differentiable. The resulting integral of a function

Math Refresher for Scientists and Engineers, Third Edition By John R. Fanchi
Copyright © 2006 John Wiley & Sons, Inc.

$f(x)$ along the curve C may be written as

$$\int_C f(x)ds$$

where x and $f(x)$ are dependent on the parameter s. This is the concept of a line integral. To see how to solve a line integral, we must be more explicit in defining the summation process and the associated path of integration (Figure 9.1).

The curve C may be written as the vector

$$\mathbf{r}(s) = x(s)\hat{i} + y(s)\hat{j} + z(s)\hat{k}$$

when the arc length s parameterizes the coordinates between the end points A, B corresponding to $s = a$ and $s = b$. The curve C, the *path of integration*, is assumed to be a smooth curve so that $\mathbf{r}(s)$ is continuous and differentiable.

The integrand $f(x, y, z)$ is a continuous function of s. As in the case of the definite integral, we form the sum

$$(I_L)_n = \sum_{m=1}^{n} f(x_m, y_m, z_m)\Delta s_m$$

where we have subdivided C into n segments of arc length

$$\Delta s_m = s_m - s_{m-1}$$

If we let $n \to \infty$ so that $\Delta s_m \to 0$, we obtain the line integral of f along the path C as

$$I_L = \int_C f(x, y, z)ds = \int_a^b f(x(s), y(s), z(s))ds$$

The value of I_L depends on the path of integration C. If the path C is closed, the integral is often written as $\oint_C f(x)dx$.

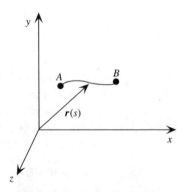

Figure 9.1   Path of integration.

Example: Evaluate the integral

$$\int_{(0,1)}^{(1,2)} (x^2 - y)\, dx + (y^2 + x)\, dy$$

along the parabola $x = t$, $y = t^2 + 1$.

The integral is evaluated by replacing x, y with their parametric representations. Since $t = 0$ at $(x, y) = (0, 1)$, and $t = 1$ at $(x, y) = (1, 2)$, the line integral becomes

$$\int_{t=0}^{1} \left\{ \left[t^2 - (t^2 + 1) \right] dt + \left[(t^2 + 1)^2 + t \right] 2t\, dt \right\}$$

$$= \int_{0}^{1} \left[2t^5 + 4t^3 + 2t^2 + 2t - 1 \right] dt$$

$$= \left[2\frac{t^6}{6} + 4\frac{t^4}{4} + 2\frac{t^3}{3} + 2\frac{t^2}{2} - t \right]_{0}^{1}$$

$$= \left[\frac{t^6}{3} + t^4 + \frac{2}{3}t^3 + t^2 - t \right]_{0}^{1} = 2$$

EXERCISE 9.1: Line Integral. Evaluate the integral $\int_C A \cdot dr$ from $(0, 0, 0)$ to $(1, 1, 1)$ along the path $x = t$, $y = t^2$, $z = t^3$ for

$$A = (3x^2 - 6yz)\hat{i} + (2y + 3xz)\hat{j} + (1 - 4xyz^2)\hat{k}$$

9.2 DOUBLE INTEGRAL

An integral of a function of two independent variables can be defined as a generalization of the definition of a definite integral. In particular, let $f(x, y)$ be defined in a closed region R (Figure 9.2). Form the sum

$$\sum_{i=1}^{n} f(x_i, y_i)\Delta A_i$$

where ΔA_i is the area of the subregion at the point (x_i, y_i). If we take the limit $n \to \infty$, we obtain the double integral of $f(x, y)$ over the region R:

$$I_D = \lim_{n \to \infty} \sum_{i=1}^{n} f(x_i, y_i)\Delta A_i = \iint_R f(x, y)\, dA$$

In Cartesian coordinates, we have

$$I_D = \iint_R f(x, y)\, dx\, dy$$

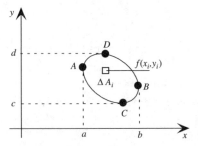

Figure 9.2   Double integral region R.

The solution of I_D requires expressing the boundary of the region R in terms of x, y. For example, suppose we write the curves ACB and ADB in Figure 9.2 as $y = g_1 (x)$ and $y = g_2 (x)$, respectively. Then I_D becomes

$$I_D = \int_a^b \left[\int_{g_1(x)}^{g_2(x)} f(x, y)\, dy \right] dx$$

and is evaluated by first integrating $f(x, y)$ over y with x fixed, then integrating over x. Similarly, suppose the curves DAC and DBC are represented by $x = h_1(y)$ and $x = h_2(y)$, respectively. The corresponding expression for the double integral is

$$I_D = \int_c^d \left[\int_{h_1(x)}^{h_2(x)} f(x, y)\, dx \right] dy$$

In this case, the integral is solved by first integrating over x with y fixed, followed by the integration over y.

If the function $f(x, y)$ has the form $f(x, y) = p(x)\, q(y)$, then the double integral can be written as the product

$$\iint_R p(x)\, q(y)\, dx\, dy = \int_a^b p(x) \left[\int_{g_1(x)}^{g_2(x)} q(y)\, dy \right] dx$$

$$= \int_c^d q(y) \left[\int_{h_1(x)}^{h_2(x)} p(x)\, dx \right] dy$$

Further simplification occurs if the limits of integration do not depend on the integration variables. Thus suppose R is defined by the inequalities $a \le x \le b, c \le y \le d$. Then

$$\int_a^b \int_c^d f(x, y)\, dy\, dx = \int_a^b p(x)\, dx \int_c^d q(y)\, dy$$

EXERCISE 9.2: Double Integral. Evaluate the double integral $\iint_R (x^2 + y^2)\, dx\, dy$ in the region R bounded by $y = x^2, x = 2, y = 1$.

9.3 FOURIER ANALYSIS

Integral transforms have many applications in science and engineering. We begin our discussion of integral transforms with the Fourier transform. The Fourier transform is an extension of the concept of the Fourier series, which we consider in this section following the introduction of two related concepts: even functions and odd functions.

Even and Odd Functions

A function $p(x)$ is said to be *even* or *symmetric* if $p(x) = p(-x)$, whereas a function $q(x)$ is called *odd* or *antisymmetric* if $q(x) = -q(-x)$. The product $r(x) = p(x)q(x)$ of an even function and an odd function is odd, since

$$r(-x) = p(-x)\, q(-x) = p(x)[-q(x)] = -r(x)$$

Example: The function $f_1(x) = x$ is odd since $f_1(-x) = -x = -f_1(x)$. The function $f_2(x) = x^2$ is even since $f_2(-x) = (-x)^2 = x^2 = f_2(x)$. The sum $f_3(x) = f_1(x) + f_2(x)$ of the two functions $f_1(x)$ and $f_2(x)$ is $f_3(x) = x + x^2$. It is neither even nor odd since $f_3(-x) = -x + (-x)^2 = -x + x^2$ does not equal either $f_3(x)$ or $-f_3(x)$. The product $f_4(x) = f_1(x)f_2(x) = x^3$ is odd since $f_4(-x) = (-x)^3 = -x^3 = -f_4(x)$.

If an even function $p(x)$ is being integrated over a symmetric interval $-a/2 \leq x \leq a/2$, where $a > 0$, then we have

$$\int_{-a/2}^{a/2} p(x)\, dx = 2 \int_{0}^{a/2} p(x)\, dx$$

since

$$\int_{-a/2}^{0} p(x)\, dx = \int_{0}^{a/2} p(x)\, dx$$

for an even function. The latter equality can be proved by making the change of variable $x \rightarrow -x$ in the integral on the left-hand side.

By contrast, if an odd function $q(x)$ is being integrated over the interval $-a/2 \leq x \leq a/2$, then we have

$$\int_{-a/2}^{a/2} q(x)\, dx = \int_{-a/2}^{0} q(x)\, dx + \int_{0}^{a/2} q(x)\, dx = 0$$

since $q(-x) = -q(x)$.

Fourier Series

The function $f(x)$ is *periodic* if it is defined for all real numbers x and satisfies the equality $f(x + nT) = f(x)$, where n is an integer and the period T is a positive number. A periodic function may be represented by a superposition, or summation, of harmonic functions. This superposition of harmonic functions is called a *Fourier expansion* of the function.

Example: The trigonometric functions $\sin x$ and $\cos x$ have period $T = 2\pi$ since $\sin(x + 2n\pi) = \sin x$ and $\cos(x + 2n\pi) = \cos x$.

The function $f(x)$ with period T and defined in the symmetric interval $-T/2 \le x \le T/2$ may be represented by the Fourier series

$$f(x) = a_0 + \sum_{n=1}^{\infty} \left[a_n \cos\left(\frac{2\pi n}{T} x\right) + b_n \sin\left(\frac{2\pi n}{T} x\right) \right] \qquad (9.3.1)$$

The Fourier coefficients $a_0, \{a_n, b_n\}$ are given by the Euler formulas:

$$a_0 = \frac{1}{T} \int_{-T/2}^{T/2} f(x) dx$$

$$a_n = \frac{2}{T} \int_{-T/2}^{T/2} f(x) \cos\left(\frac{2\pi n}{T} x\right) dx \qquad (9.3.2)$$

$$b_n = \frac{2}{T} \int_{-T/2}^{T/2} f(x) \sin\left(\frac{2\pi n}{T} x\right) dx$$

A perusal of the equation for the coefficient a_0 shows that a_0 is the mean or average value of $f(x)$ over the period T.

Example: The Fourier series of $f(x) = x$ in the interval $-\pi \le x \le \pi$ is

$$f(x) = \sum_{n=1}^{\infty} b_n \sin nx, \ b_n = \frac{2}{n}(-1)^{n+1}$$

The Fourier coefficients a_0 and $\{a_n\}$ are zero because they are calculated by integrating an odd function over the symmetric interval $-\pi \le x \le \pi$. The $\{b_n\}$ terms are obtained by an integration by parts, that is,

$$b_n = \frac{2}{2\pi} \int_{-\pi}^{\pi} x \sin nx \, dx = \frac{1}{\pi} \left\{ -\left. \frac{x \cos nx}{n} \right|_{-\pi}^{\pi} + \frac{1}{n} \int_{-\pi}^{\pi} \cos nx \, dx \right\}$$

$$= \frac{1}{\pi} \left\{ -\frac{2\pi}{n} \cos n\pi + \frac{1}{n^2} \sin nx \Big|_{-\pi}^{\pi} \right\} = \frac{2}{n}(-1)^{n+1}$$

Some frequently encountered integrals for a period $T = 2\pi$ are summarized below. Notice that the Kronecker delta $\delta_{mn} = 0$ when $m \neq n$ and $\delta_{mn} = 1$ when $m = n$.

$$\frac{1}{\pi} \int_{-\pi}^{\pi} \sin nt \sin mt \, dt = \delta_{mn}$$

$$\frac{1}{\pi} \int_{-\pi}^{\pi} \cos nt \cos mt \, dt = \delta_{mn}$$

$$\frac{1}{\pi} \int_{-\pi}^{\pi} \cos nt \sin mt \, dt = 0$$

A set of functions $\{\phi_n(t): n = 1, 2, 3, \dots\}$ that has the property

$$\int_a^b \phi_m(t)\phi_n(t) \, dt = \delta_{mn}$$

is called an *orthonormal* set. The function $\phi_m(t)$ is orthogonal to $\phi_n(t)$ if $m \neq n$, and the function is square integrable with unit normalization if $m = n$, that is,

$$\int_a^b \left[\phi_m(t)\right]^2 dt = 1$$

EXERCISE 9.3: Show that $\left\{\phi_n(x) = (\sin nx)/\sqrt{\pi} : 1, 2, 3, \dots\right\}$ is an orthonormal set of functions in the interval $-\pi \leq x \leq \pi$.

Notice that the integrals for the coefficients $a_0, \{a_n\}$ in Eq. (9.3.2) vanish when the function $f(x)$ with period T is odd, and the coefficients $\{b_n\}$ in Eq. (9.3.2) are zero when $f(x)$ is even. Thus the Fourier series for an even function $p(x)$ with period T is

$$p(x) = a_0 + \sum_{n=1}^{\infty} a_n \cos\left(\frac{2\pi n}{T}x\right)$$

with coefficients

$$a_0 = \frac{2}{T} \int_0^{T/2} p(x) \, dx$$

$$a_n = \frac{4}{T} \int_0^{T/2} p(x) \cos\left(\frac{2\pi n}{T}x\right) dx$$

Similarly, the Fourier series for an odd function $q(x)$ with period T is

$$q(x) = \sum_{n=1}^{\infty} b_n \sin\left(\frac{2\pi n}{T}x\right)$$

with coefficients

$$b_n = \frac{4}{T} \int_0^{T/2} q(x) \sin\left(\frac{2\pi n}{T} x\right) dx$$

The Fourier series of a function $f(x)$ of a real variable x may be written in complex form as

$$f(x) = \sum_{n=-\infty}^{\infty} c_n e^{inx}$$

with complex coefficients

$$c_n = \frac{1}{2\pi} \int_{-\pi}^{\pi} f(x) e^{-inx} dx \text{ for } n = 0, \ \pm 1, \ \pm 2, \ldots$$

The Fourier series of a function $f(z)$ of a complex variable z can be derived using Euler's equation $e^{i\theta} = \cos\theta + i\sin\theta$.

9.4 FOURIER INTEGRAL AND FOURIER TRANSFORM

A Fourier series expansion of a function $f(x)$ of a real variable x with period T is defined over a finite interval $-T/2 \leq x \leq T/2$. If the interval becomes infinite and we sum over infinitesimals, we obtain the *Fourier integral*

$$f(x) = \frac{1}{(2\pi)^{1/2}} \int_{-\infty}^{\infty} A(k) e^{ikx} dk \qquad (9.4.1)$$

with the coefficients

$$A(k) = \frac{1}{(2\pi)^{1/2}} \int_{-\infty}^{\infty} f(x) e^{-ikx} dx \qquad (9.4.2)$$

Equation (9.4.2) is the *Fourier transform* of $f(x)$. The Fourier integral is also known as the *inverse Fourier transform* of $A(k)$.

Example: Suppose we have a square waveform with constant pulse height H defined by

$$f(x) = \begin{cases} H, \ -a \leq x \leq a \\ 0, \ |x| > a \end{cases}$$

Then the Fourier transform is

$$A(k) = \frac{1}{(2\pi)^{1/2}} \left\{ \int_{-\infty}^{-a} 0 \cdot e^{-ikx} dx + \int_{-a}^{a} H \cdot e^{-ikx} dx + \int_{a}^{\infty} 0 \cdot e^{-ikx} dx \right\}$$

$$= \frac{H}{(2\pi)^{1/2}} \int_{-a}^{a} e^{-ikx} dx = \left[\frac{H}{\sqrt{2\pi}} \frac{e^{-ikx}}{-ik} \right]_{-a}^{a}$$

$$= \frac{H}{\sqrt{2(\pi)}} \frac{e^{+ika} - e^{-ika}}{ik} = \frac{Ha}{\sqrt{2(\pi)}} \frac{\sin ka}{ka}$$

The function $(\sin \theta)/\theta$ appears often enough in Fourier transform theory that it has been given the special name sinc θ and is defined as

$$\text{sinc } \theta = \frac{\sin \theta}{\theta}$$

In this notation, the Fourier transform of a square waveform is

$$A(k) = \frac{Ha}{\sqrt{2\pi}} \text{sinc } ka$$

and its Fourier integral is

$$f(x) = \frac{Ha}{2\pi} \int_{-\infty}^{\infty} \frac{\sin ka}{ka} e^{ikx} dk$$

EXERCISE 9.4: Determine the Fourier transform of the harmonic function

$$g(t) = \begin{cases} \exp(i\omega_0 t), & -T \le t \le T \\ 0, & |t| > 0 \end{cases}$$

Dirac Delta Function

If we substitute the Fourier transform $A(k)$ into the Fourier integral for $f(x)$, we obtain

$$f(x) = \frac{1}{2\pi} \int_{-\infty}^{\infty} \left[\int_{-\infty}^{\infty} e^{-iky} f(y) \, dy \right] e^{ikx} dk$$

The double integral can be rearranged to give

$$f(x) = \int_{-\infty}^{\infty} \delta(x - y) f(y) \, dy \qquad (9.4.3)$$

where the mathematical distribution

$$\delta(x - y) \equiv \frac{1}{2\pi} \int_{-\infty}^{\infty} e^{ik(x-y)} dk \tag{9.4.4}$$

is called the *Dirac delta function*. The Dirac delta function is well defined only within the context of integration. It is widely used as a mathematical representation of a point source or sink.

Alternative Forms of the Fourier Integral and Transform

The notation in Eqs. (9.4.1) and (9.4.2) implies that the Fourier integral $f(x)$ and Fourier transform $A(k)$ are defined in the space (x) and wavenumber (k) domains, respectively. In the time (t) and frequency (ω) domains, Eqs. (9.4.1) and (9.4.2) have the forms

$$f(t) = \frac{1}{(2\pi)^{1/2}} \int_{-\infty}^{\infty} A(\omega) e^{i\omega t} d\omega \tag{9.4.5}$$

and

$$A(\omega) = \frac{1}{(2\pi)^{1/2}} \int_{-\infty}^{\infty} f(t) e^{-i\omega t} dt \tag{9.4.6}$$

Suppose the Fourier transform $A(\omega)$ is written as the product

$$A(\omega) = \Phi(\omega) e^{i\phi(\omega)} \tag{9.4.7}$$

The term $\Phi(\omega)$ is called the *frequency spectrum* of $f(t)$ and $\phi(\omega)$ is called the *phase spectrum* of $f(t)$.

The placement of the factor $(2\pi)^{-1/2}$ in Eqs. (9.4.1), (9.4.2), (9.4.5), and (9.4.6) is a matter of choice as long as the Fourier integral

$$f(t) = \frac{1}{(2\pi)} \int_{-\infty}^{\infty} \left[\int_{-\infty}^{\infty} f(s) e^{-i\omega s} ds \right] e^{i\omega t} d\omega \tag{9.4.8}$$

is satisfied. The variable s in Eq. (9.4.8) is a dummy variable of integration. An alternative definition of the Fourier integral is

$$g(t) = \frac{1}{2\pi} \int_{-\infty}^{\infty} B(\omega) e^{i\omega t} d\omega \tag{9.4.9}$$

with the corresponding Fourier transform

$$B(\omega) = \int_{-\infty}^{\infty} g(t) e^{-i\omega t} dt \tag{9.4.10}$$

Equation (9.4.10) is the inverse Fourier transform of $B(\omega)$. The Fourier transform in Eq. (9.4.10) may be written in the notation

$$\mathcal{F}\{g(t)\} = B(\omega) = \int_{-\infty}^{\infty} g(t)e^{-i\omega t}\,dt \qquad (9.4.11)$$

In this notation, the inverse Fourier transform in Eq. (9.4.9) is

$$\mathcal{F}^{-1}\{B(\omega)\} = g(t) = \frac{1}{2\pi}\int_{-\infty}^{\infty} B(\omega)e^{i\omega t}\,d\omega \qquad (9.4.12)$$

EXERCISE 9.5: Show that Eq. (9.4.8) is valid. *Hint:* You should use the Dirac delta function in Eq. (9.4.4).

Convolution Integral

Consider a double integral of two functions whose variables of integration x,y are constrained by a delta function $\delta(K - x - y)$ such that

$$I_{con} = \iint f(x)\,g(y)\,\delta(K - x - y)\,dx\,dy$$

The double integral can be reduced to a single integral by integrating over the delta function to find

$$I_{con} = \int f(x)\,g(K - x)\,dx = \int f(K - y)\,g(y)\,dy \qquad (9.4.13)$$

The integrals in Eq. (9.4.13) are known as *convolution integrals* and are often written in the form

$$I_{con} = f(x) \circ g(K - x) = f(K - y) \circ g(y) \qquad (9.4.14)$$

EXERCISE 9.6: Show that the Fourier transform of the product of two functions $g_1(t)$ and $g_2(t)$ can be written as a convolution integral. *Hint:* You should use Eqs. (9.4.11) and (9.4.12) and the Dirac delta function in Eq. (9.4.4).

9.5 TIME SERIES AND Z TRANSFORM

Time series and Z transforms are used in a variety of real-world applications, such as signal processing and seismic analysis [Collins, 1999; Yilmaz, 2001; Telford et al., 1990]. The concepts of time series and Z transforms are reviewed here, as is the relationship between the Z transform and the Fourier transform.

Time Series and Digital Functions

Consider a function $f(t)$ that is a continuous and single-valued function of time t. Time is viewed here as a continuous variable. A *time series* is a series of values that is obtained by sampling the function $f(t)$ at regular time intervals. The resulting time series is a sequence with each value in the sequence corresponding to a discrete value of t.

Z Transforms

Telford et al. [1990] call the sequence corresponding to a time series a *digital function*. For example, a digital function h_t may be written as the sequence

$$h_t = \{h_n\} = \{\ldots, h_{-2}, h_{-1}, h_0, h_1, h_2, \ldots\} \tag{9.5.1}$$

where n is an integer in the range $-\infty < n < \infty$. The value of the continuous variable t that corresponds to n is

$$t = n\,\Delta t \tag{9.5.2}$$

where Δt is a constant increment. The digital function h_t is a *causal digital function* if $\{h_n = 0\}$ for all $n < 0$. Define the Z transform of a digital function $g_t = \{g_n\}$ for the sequence $\{g_n\} = \{\ldots, g_{-2}, g_{-1}, g_0, g_1, g_2, \ldots\}$ as

$$g(z) = \sum_{n=-\infty}^{\infty} g_n z^n \tag{9.5.3}$$

If $h_t = \{h_n\}$ is a causal digital function, the elements of the sequence $h_t = \{h_n\}$ for all $n < 0$ are zero and the Z transform is

$$h(z) = \sum_{n=0}^{\infty} h_n z^n \tag{9.5.4}$$

The Z transform can be used to represent a sequence that is obtained by sampling a continuous function.

EXERCISE 9.7: What are the Z transforms of the causal digital functions $h_1 = \{1, -2, -1, 0, 1\}$ and $h_2 = \{0.6, -1.2, -2.4, 0.7, 3.8\}$?

Alternative Forms of the Z Transform

The definition of the Z transform given in Eq. (9.5.3) is one of two definitions in the literature. We can see the difference between alternative forms of the Z transform by considering the relationship between the Z transform and the Fourier transform.

Consider a function of the form

$$g(t) = \sum_{n=-\infty}^{\infty} g_n \delta(t - n\,\Delta t) \tag{9.5.5}$$

where $\delta(t - n\,\Delta t)$ is a Dirac delta function and Δt is a constant increment. Substituting Eq. (9.5.5) into the Fourier transform given by Eq. (9.4.12) gives

$$B(\omega) = \int_{-\infty}^{\infty} \left[\sum_{n=-\infty}^{\infty} g_n \delta(t - n\,\Delta t) \right] e^{-i\omega t} dt \tag{9.5.6}$$

Interchanging the sum and integral lets us write

$$B(\omega) = \sum_{n=-\infty}^{\infty} \left[\int_{-\infty}^{\infty} g_n \delta(t - n\,\Delta t) e^{-i\omega t} dt \right] \tag{9.5.7}$$

Evaluating the integral gives

$$B(\omega) = \sum_{n=-\infty}^{\infty} g_n e^{-i\omega n\,\Delta t} \tag{9.5.8}$$

Two forms of the Z transform can be obtained from Eq. (9.5.8).

If we define the variable

$$z_1 = e^{-i\omega\,\Delta t} \tag{9.5.9}$$

and substitute into Eq. (9.5.8), we get

$$B(z_1) = \sum_{n=-\infty}^{\infty} g_n z_1^n \tag{9.5.10}$$

Equation (9.5.10) is the definition of the Z transform used in seismic analysis [Yilmaz, 2001; Telford et al., 1990]. By contrast, if we define the variable

$$z_2 = e^{i\omega\,\Delta t} \tag{9.5.11}$$

and substitute into Eq. (9.5.8), we get

$$B(z_2) = \sum_{n=-\infty}^{\infty} g_n z_2^{-n} \tag{9.5.12}$$

Equation (9.5.12) is the definition of the Z transform used by authors such as Zwillinger [2003] and Collins [1999]. If we make the variable substitution

$$s = z_2^{-1} \tag{9.5.13}$$

in Eq. (9.5.11), we obtain

$$B(s) = \sum_{n=-\infty}^{\infty} g_n s^n \tag{9.5.14}$$

Equation (9.5.14) is the generating function of the sequence $\{g_n\}$ [Zwillinger, 2003].

9.6 LAPLACE TRANSFORM

The Laplace transform of a function $f(x)$ of a variable x is defined as the integral

$$F(s) = \mathcal{L}\{f(x)\} = \int_0^\infty e^{-sx} f(x)\, dx \qquad (9.6.1)$$

where s is a real, positive parameter that serves as a supplementary variable. The Laplace transform of the sum or difference of two functions $f(x)$, $g(x)$ is

$$\mathcal{L}\{f(x) \pm g(x)\} = \int_0^\infty e^{-sx}[f(x) \pm g(x)]dx$$

$$= \int_0^\infty e^{-sx} f(x)\ dx \pm \int_0^\infty e^{-sx} g(x)\, dx$$

Using the definition of the Laplace transform, we find

$$\mathcal{L}\{f(x) \pm g(x)\} = \mathcal{L}\{f(x)\} \pm \mathcal{L}\{g(x)\} \qquad (9.6.2)$$

The inverse of the Laplace transform is

$$\mathcal{L}^{-1}\{F(s)\} = f(x) \qquad (9.6.3)$$

Example: The Laplace transform of $f(x) = e^{ax}$ is

$$F(s) = \mathcal{L}\{e^{ax}\} = \int_0^\infty e^{-sx} e^{ax} dx = \int_0^\infty e^{-(s-a)x} dx$$

This integral is called an *improper integral* because one of the integration limits is ∞. To evaluate the improper integral we assume $s > a$ to find

$$\mathcal{L}\{e^{ax}\} = \left[\frac{e^{-(s-a)x}}{s - a}\right]_0^\infty = \frac{1}{s - a}$$

Several examples of Laplace transforms are given below, where a is a constant and n is an integer.

LAPLACE TRANSFORMS

$f(x) = \mathcal{L}^{-1}\{F(s)\}$	$F(s) = \mathcal{L}\{f(x)\}$
a	a/s
x^n	$(n!)/s^{n+1}$
e^{ax}	$1/(s - a)$
$\sin ax$	$a/(s^2 + a^2)$
$\cos ax$	$s/(s^2 + a^2)$
$\sinh ax$	$a/(s^2 - a^2)$
$\cosh ax$	$s/(s^2 - a^2)$

Laplace Transforms of Derivatives

The Laplace transform of the first derivative $f'(x)$ is

$$\mathcal{L}\left\{\frac{df(x)}{dx}\right\} = \int_0^\infty e^{-sx} \frac{df(x)}{dx} dx$$

Performing an integration by parts of the form $\int u \, dv = uv - \int v \, du$ gives

$$\mathcal{L}\{f'(x)\} = [e^{-sx} f(x)]_0^\infty + s \int_0^\infty e^{-sx} f(x) \, dx$$

We assume $e^{-sx} f(x) \to 0$ as $x \to \infty$ so that

$$\mathcal{L}\{f'(x)\} = -f(0) + s\mathcal{L}\{f(x)\} \tag{9.6.4}$$

where we have used the definition of Laplace transform.

The Laplace transform of $f'(x)$ can be used to calculate higher-order derivatives. For example, to find the Laplace transform of the second-order derivative $\mathcal{L}\{f''(x)\}$, let $g(x) = f'(x)$ so that $g'(x) = f''(x)$. Then we have by repeated application of $\mathcal{L}\{f'(x)\}$ the result

$$\mathcal{L}\{g'(x)\} = s\mathcal{L}\{g(x)\} - g(0) = s[s\mathcal{L}\{f(x)\} - f(0)] - f'(0)$$

or

$$\mathcal{L}\{f''(x)\} = s^2 \mathcal{L}\{f(x)\} - sf(0) - f'(0) \tag{9.6.5}$$

Laplace Transform of an Integral

The Laplace transform of an integral

$$h(t) = \int_0^t i(\tau) \, d\tau \tag{9.6.6}$$

is found by first writing the Laplace transform of the functions $i(t)$ and $h(t)$ as

$$\mathcal{L}\{i(t)\} = I(s) \tag{9.6.7}$$

and

$$\mathcal{L}\{h(t)\} = H(s) \tag{9.6.8}$$

Equation (9.6.6) is equivalent to

$$\frac{dh(t)}{dt} = i(t) \tag{9.6.9}$$

given the initial condition $h(0) = 0$ corresponding to the lower limit of the integral in Eq. (9.6.6). The Laplace transform of Eq. (9.6.9) given $h(0) = 0$ is found from Eq. (9.6.4) to be

$$\mathcal{L}\left\{\frac{dh(t)}{dt}\right\} = sH(s) \qquad (9.6.10)$$

We rearrange Eq. (9.6.10) and use Eqs. (9.6.7) and (9.6.8) to find

$$H(s) = \frac{\mathcal{L}\{dh(t)/dt\}}{s} = \frac{\mathcal{L}\{i(t)\}}{s} = \frac{I(s)}{s} \qquad (9.6.11)$$

Substituting Eq. (9.6.6) into Eq. (9.6.8) and using the result in Eq. (9.6.11) gives the Laplace transform of the integral in Eq. (9.6.6) as

$$\mathcal{L}\left\{\int_0^t i(\tau)\,d\tau\right\} = \frac{I(s)}{s} = \frac{\mathcal{L}\{i(t)\}}{s} \qquad (9.6.12)$$

EXERCISE 9.8: Verify the Laplace transform $\mathcal{L}\{af(t)\} = a\mathcal{L}\{f(t)\}$, where a is a complex constant, and $f(t)$ is a function of the variable t.

EXERCISE 9.9: Verify the inverse Laplace transform $\mathcal{L}^{-1}\{aF(s)\} = a\mathcal{L}^{-1}\{F(s)\}$, where a is a complex constant, and $F(s) = \mathcal{L}\{f(t)\}$ is the Laplace transform of a function $f(t)$ of the variable t.

EXERCISE 9.10: What is the Laplace transform of the complex function $e^{i\omega t}$ with variable t and constant ω? Use the Laplace transform of $e^{i\omega t}$ and Euler's equation $e^{i\omega t} = \cos \omega t + i \sin \omega t$ to verify the Laplace transforms $\mathcal{L}\{\cos \omega t\} = s/(s^2 + \omega^2)$ and $\mathcal{L}\{\sin \omega t\} = \omega/(s^2 + \omega^2)$.

CHAPTER 10

ORDINARY DIFFERENTIAL EQUATIONS

Mathematical models of natural phenomena usually include terms for describing the change of one or more variables resulting from the change of one or more other variables. If the equation describing such a system has the form $F(x, y(x), dy(x)/dy(x)/dx, \ldots, d^m y(x)/dx^m) \equiv F(\cdot) = 0$, where x is the single independent variable and y is a function of x, then $F(\cdot) = 0$ is an mth order ordinary differential equation (ODE) with solution $y(x)$. In this chapter we review the solution of first-order ODEs and a procedure for converting higher-order ODEs to systems of first-order ODEs. The latter procedure makes it possible to study the stability of the original ODE without actually finding a solution. Several ODE solution techniques are presented in Chapter 11.

10.1 FIRST-ORDER ODE

A linear, first-order ODE has the form

$$\frac{dy}{dx} + f(x)y = r(x) \tag{10.1.1}$$

where x is an independent variable, y is a dependent variable, and $f(x)$ and $r(x)$ are arbitrary functions of x. Equation (10.1.1) is linear because the variable y appears only with a power of 1, and it is first-order because the highest-order derivative is

Math Refresher for Scientists and Engineers, Third Edition By John R. Fanchi
Copyright © 2006 John Wiley & Sons, Inc.

first-order. The order of the ODE is determined by the highest-order derivative. Two special cases depend on the value of $r(x)$. In particular, Eq. (10.1.1) is homogeneous if $r(x) = 0$, and Eq. (10.1.1) is nonhomogeneous if $r(x) \neq 0$.

Quadrature Solution

Consider the homogeneous equation

$$\frac{dy_h}{dx} + f(x)y_h = 0 \qquad (10.1.2)$$

We separate variables by collecting explicit functions of x and y on opposite sides of the equation to get

$$\frac{dy_h}{y_h} = -f(x)dx$$

Integrating gives

$$\ln y_h = -\int f(x)dx + c_1$$

where c_1 is the integration constant. The formal solution of the linear, first-order, homogeneous ODE is

$$y_h = c_2 e^{-\int f(x)dx}, c_2 = \pm e^{c_1} \qquad (10.1.3)$$

Equation (10.1.3) is called the quadrature solution because it is expressed in terms of the quadrature, or integral, of $f(x)$. The final solution of the ODE requires solving the integral in Eq. (10.1.3).

Example: Suppose $f(x) = x$ and we want to find the solution of the homogeneous ODE

$$\frac{dy_h}{dx} + xy_h = 0$$

By Eq. (10.1.3) we have

$$y_h = c_2 e^{-\int f(x)dx} = c_2 e^{-\int x dx}$$

Performing the integration gives

$$y_h = c_2 e^{-(x^2/2)+c_3} = c_2 e^{c_3} e^{-x^2/2}$$

where c_3 is the integration constant. The term $c_2 e^{c_3}$ is a constant term because the three factors e, c_2, c_3 are constants. The term $c_2 e^{c_3}$ can be replaced by a new constant, c_4. Using c_4 lets us simplify the solution; thus

$$y_h = c_4 e^{-x^2/2}$$

To verify our solution, we calculate the derivative

$$\frac{dy_h}{dx} = c_4 e^{-x^2/2} \left[\frac{d(-x^2/2)}{dx} \right] = -\frac{1}{2} y_h(2x) = -xy_h$$

This reproduces our original homogeneous ODE.

Let us now consider the nonhomogeneous form of Eq. (10.1.1). To solve Eq. (10.1.1), we first multiply by e^h, where

$$h = \int f(x)dx$$

to get

$$e^h \frac{dy}{dx} + e^h fy = e^h r$$

The new ODE can be rearranged by noting that the left-hand side has the form

$$\frac{d}{dx}(e^h y) = y \frac{de^h}{dx} + e^h \frac{dy}{dx}$$

$$= ye^h \frac{dh}{dx} + e^h \frac{dy}{dx} = yfe^h + e^h \frac{dy}{dx}$$

(10.1.4)

Therefore Eq. (10.1.4) becomes

$$\frac{d}{dx}(e^h y) = re^h$$

Separating variables gives a relationship between differentials:

$$d(e^h y) = re^h \, dx$$

Upon integration we find

$$ye^h = \int e^h r \, dx + C$$

where C is the constant of integration. Rearranging gives the general solution to the nonhomogeneous, linear, first-order ODE as

$$y(x) = e^{-h} \left[\int e^h r \, dx + C \right], h = \int f(x) \, dx$$

(10.1.5)

EXERCISE 10.1: Solve

$$\frac{dy}{dx} - y = e^{2x}$$

In general, an mth order ODE has m unspecified constants in the solution. These constants are determined by specifying m additional constraints on the solution. The constraint imposed on the solution of a first-order ODE is usually called the initial condition if the independent variable is time. If the independent variable is a space coordinate, the constraint is called a boundary condition. An example of an initial value problem is a first-order ODE with the specified initial condition

$$y(x_0) = y_0$$

EXERCISE 10.2: Solve

$$\frac{dy}{dx} + y \tan x = \sin 2x, y(0) = 1$$

Systems of First-Order ODEs

Consider the following system of equations:

$$\frac{dx_1}{dt} = f_1(x_1, \ldots, x_n, t)$$

$$\frac{dx_2}{dt} = f_2(x_1, \ldots, x_n, t)$$

$$\vdots$$

$$\frac{dx_n}{dt} = f_n(x_1, \ldots, x_n, t)$$

where the variables $\{x_i : i = 1, \ldots, n\}$ depend on the independent variable t. This set of equations may be written in matrix form as

$$\frac{d}{dt}x = f(x, t) \tag{10.1.6}$$

Equation (10.1.6) represents a system of equations that are called nonautonomous because t appears explicitly in f. An autonomous system has the form

$$\frac{d}{dt}y = f(y) \tag{10.1.7}$$

If an initial vector $x(t_0) = x^0$ or $y(t_0) = y^0$ is specified, then we have an initial value problem.

A nonautonomous, initial value problem can be made into an autonomous problem by adding the first-order ODE

$$\frac{dx_{n+1}}{dt} = 1, x_{n+1}(t = 0) = t_0 \qquad (10.1.8)$$

with the initial condition as shown. Equation (10.1.8) has the solution

$$x_{n+1}(t) = t + t_0$$

and Eq. (10.1.6) becomes

$$\frac{d}{dt}x_A = f_A(x_A)$$

where

$$x_A = \begin{bmatrix} x \\ t \end{bmatrix}, f_A = \begin{bmatrix} f(x_A) \\ 1 \end{bmatrix}$$

are $n+1$ column vectors.

EXERCISE 10.3: Convert the nonautonomous equation

$$\frac{dx_1}{dt} = 1 - t + 4x_1, x_1(0) = 1$$

to an autonomous system.

Runge–Kutta Method

Systems of linear ODEs may be solved numerically using techniques such as the Runge–Kutta fourth-order numerical algorithm [Press et al., 1992, Sec. 16.1; Edwards and Penney, 1989, Sec. 6.4]. Suppose the initial conditions are

$$x(t_0) = x_0$$

at $t = t_0$ for the system of equations

$$\frac{d}{dt}x = f(x, t)$$

Values of x as functions of t are found by incrementally stepping forward in t. The fourth-order Runge–Kutta method calculates new values of x_{n+1} from old values x_n using the algorithm

$$x_{n+1} = x_n + \frac{h}{6}[w_1 + 2w_2 + 2w_3 + w_4] + O(h^5)$$

where h is an incremental step size $0 < h < 1$. The terms of the algorithm are

$$t_{n+1} = t_n + h$$
$$w_1 = f(x_n, t_n)$$
$$w_2 = f\left(x_n + \frac{1}{2}hw_1, t_n + \frac{1}{2}h\right)$$
$$w_3 = f\left(x_n + \frac{1}{2}hw_2, t_n + \frac{1}{2}h\right)$$
$$w_4 = f(x_n + hw_3, t_n + h)$$

The calculation begins at $n = 0$ and proceeds iteratively. At the end of each step, the new values are defined as present values at the nth level and another iteration is performed.

Example: Suppose we want to solve a system of two first-order ODEs of the form

$$\frac{dx_1}{dt} = x_2$$
$$\frac{dx_2}{dt} = -x_1$$

(10.1.9)

with initial conditions

$$x(t_0) = \begin{bmatrix} x_{10} \\ x_{20} \end{bmatrix} = \begin{bmatrix} 0 \\ 1 \end{bmatrix}$$

(10.1.10)

The column vectors x and f are given by

$$x = \begin{bmatrix} x_1 \\ x_2 \end{bmatrix}, f = \begin{bmatrix} f_1 \\ f_2 \end{bmatrix} = \begin{bmatrix} x_2 \\ -x_1 \end{bmatrix}$$

The matrix equation to be solved for x_{n+1} given x_n and t is

$$x_{n+1} = x_n + \frac{h}{6}(a + 2b + 2c + d)$$

where $t_{n+1} = t_n + h$ and

$$a = f(x_n, t_n)$$
$$b = f\left(x_n + \frac{h}{2}a, t_n + \frac{h}{2}\right)$$
$$c = f\left(x_n + \frac{h}{2}b, t_n + \frac{h}{2}\right)$$
$$d = f(x_n + hc, t_n + h)$$

A FORTRAN 90/95 program for solving this problem is as follows:

RUNGE-KUTTA ALGORITHM IN FORTRAN 90/95

```fortran
! RK_4th - 4TH ORDER RUNGE-KUTTA ALGORITHM FOR A SYSTEM
! WITH TWO 1ST ORDER DIFFERENTIAL EQUATIONS
! J.R. FANCHI, MAY 2003
!
! DEFINE RHS FUNCTIONS
!
  FN1(X1,X2,T)=X2
  FN2(X1,X2,T)=-X1
!
! INITIAL CONDITIONS AND R-K ALGORITHM CONTROL
! PARAMETERS
!
  DATA T0,X10,X20/0,0,1/
  DATA H,NS/0.1,20/
!
! INITIALIZE VARIABLES
!
  X1=X10
  X2=X20
  T=T0
  OPEN(UNIT=6,FILE='RK_4th.OUT')
  WRITE(6,10)
10 FORMAT(5X,18X,'NUMERICAL',10X,'ANALYTICAL', &/2X, &
  'STEP T X1 X2 X1AN X2AN')
!
! R-K ITERATION
!
  DO I=1,NS
  A1=FN1(X1,X2,T)
  A2=FN2(X1,X2,T)
  B1=FN1(X1+(H*A1/2),X2+(H*A2/2),T+(H/2))
  B2=FN2(X1+(H*A1/2),X2+(H*A2/2),T+(H/2))
  C1=FN1(X1+(H*B1/2),X2+(H*B2/2),T+(H/2))
  C2=FN2(X1+(H*B1/2),X2+(H*B2/2),T+(H/2))
  D1=FN1(X1+(H*C1),X2+(H*C2),T+H)
  D2=FN2(X1+(H*C1),X2+(H*C2),T+H)
! UPDATE X,Y,T
  X1=X1+(H/6)*(A1+2*B1+2*C1+D1)
  X2=X2+(H/6)*(A2+2*B2+2*C2+D2)
  T=T+H
```

```
! ANALYTICAL SOLUTION
  X1AN=SIN(T)
  X2AN=COS(T)
! OUTPUT RESULTS
  WRITE(6,20) I,T,X1,X2,X1AN,X2AN
20 FORMAT(1X,I5,5(2X,F8.6))
  ENDDO
!
  STOP
  END
```

Given the initial conditions in Eq. (10.1.10), the system in Eq. (10.1.9) has the analytic solution $x_1(t) = \sin t$ and $x_2(t) = \cos t$. A comparison of analytical and numerical results is shown as follows for $h = 0.1$. Notice the agreement between the analytical and numerical results. For more discussion of this example, see Edwards and Penney [1989, Sec. 6.4]. Additional examples are presented in texts such as Chapra and Canale [2002], Kreyszig [1999], and DeVries [1994].

| | Numerical | | Analytical | |
| | x_1 | x_2 | x_1 | x_2 |
t				
0.100000	0.099833	0.995004	0.099833	0.995004
0.200000	0.198669	0.980067	0.198669	0.980067
0.300000	0.295520	0.955337	0.295520	0.955337
0.400000	0.389418	0.921061	0.389418	0.921061
0.500000	0.479425	0.877583	0.479426	0.877583
0.600000	0.564642	0.825336	0.564642	0.825336
0.700000	0.644217	0.764843	0.644218	0.764842
0.800000	0.717356	0.696707	0.717356	0.696707
0.900000	0.783326	0.621611	0.783327	0.621610
1.000000	0.841470	0.540303	0.841471	0.540302
1.100000	0.891207	0.453597	0.891207	0.453596
1.200000	0.932039	0.362359	0.932039	0.362358
1.300000	0.963558	0.267500	0.963558	0.267499
1.400000	0.985449	0.169968	0.985450	0.169967
1.500000	0.997495	0.070738	0.997495	0.070737
1.600000	0.999574	−0.029198	0.999574	−0.029200
1.700000	0.991665	−0.128843	0.991665	−0.128845
1.800000	0.973848	−0.227201	0.973848	−0.227202
1.900000	0.946301	−0.323288	0.946300	−0.323290
2.000000	0.909298	−0.416145	0.909297	−0.416147

The fourth-order Runge–Kutta method is a robust method; that is, it is capable of solving a wide range of problems, but it is not necessarily the most accurate method for all applications. Predictor-corrector methods, for example, are often used in astrophysics. Specialized numerical methods references should be consulted for further details.

EXERCISE 10.4: Convert the FORTRAN 90/95 program for the Runge–Kutta algorithm in the above example to the C++ programming language. Show that the numerical and analytical results agree.

10.2 HIGHER-ORDER ODE

An nth order ODE may be written in the form

$$\frac{d^n y}{dt^n} = F\left(y, \frac{dy}{dt}, \ldots, \frac{d^{n-1}y}{dt^{n-1}}\right)$$

Suppose the initial conditions have the form

$$y(t_0), \left.\frac{dy}{dt}\right|_{t_0}, \ldots, \left.\frac{d^{n-1}y}{dt^{n-1}}\right|_{t_0}$$

If we introduce the variables

$$x_1(t) = y, \, x_2(t) = \frac{dy}{dt}, \ldots, x_n(t) = \frac{d^{n-1}y}{dt^{n-1}}$$

we can write the nth order ODE as a system of first-order ODEs such that

$$\frac{dx_1}{dt} = x_2$$

$$\frac{dx_2}{dt} = x_3$$

$$\vdots$$

$$\frac{dx_n}{dt} = F(x_1, \ldots, x_n)$$

The corresponding initial conditions are

$$x_1(t_0) = y(t_0), \, x_2(t_0) = \left.\frac{dy}{dt}\right|_{t_0}, \ldots, x_n(t_0) = \left.\frac{d^{n-1}y}{dt^{n-1}}\right|_{t_0}$$

EXERCISE 10.5: Write the third-order ODE

$$\frac{d^3y}{dt^3} = 2y\left(\frac{dy}{dt}\right)^2 + 4y$$

as a first-order system.

EXERCISE 10.6: The linear harmonic oscillator is governed by

$$\frac{d^2y}{dt^2} + y = 0, y(0) = a, \left.\frac{dy}{dt}\right|_0 = b$$

Convert this second-order ODE to a first-order system.

One practical advantage of transforming an nth order ODE into a system of first-order ODEs is that the methods for solving the first-order system become applicable to solving the nth order ODE. Thus a numerical method like the Runge–Kutta method can be used to solve an nth order ODE after it has been transformed into a system of first-order ODEs. The Runge–Kutta example presented in Section 10.1 illustrates this technique for the second-order ODE $y'' + y = 0$ and initial conditions $y(0) = 0$, $y'(0) = 1$. For more details, see Edwards and Penney [1989, Sec. 6.4].

10.3 STABILITY ANALYSIS

Many realistic models of physical systems require mathematics that are intractable, yet we still would like information about the system. One of the most important pieces of information of interest to us is the stability of a dynamical system. A dynamical system is a system that changes with time t. We show here how to analyze the stability of a dynamical system [Beltrami, 1987] described by the second-order differential equation

$$\frac{d^2\Lambda}{dt^2} + \gamma_1 \frac{d\Lambda}{dt} + \gamma_0\Lambda = 0 \tag{10.3.1}$$

The coefficients $\{\gamma_i\}$ may be arbitrary functions of t.

The second-order differential equation is transformed to a set of first-order differential equations by defining

$$\chi_1 = \Lambda, \chi_2 = \frac{d\Lambda}{dt}$$

The corresponding set of first-order differential equations is

$$\frac{d\chi_1}{dt} = \chi_2 \equiv f_1, \frac{d\chi_2}{dt} = -\gamma_0\chi_1 - \gamma_1\chi_2 \equiv f_2 \tag{10.3.2}$$

Equation (10.3.2) can be cast in the matrix form

$$\dot{\chi} = A\chi \equiv f \tag{10.3.3}$$

where

$$\dot{\chi} = \frac{d}{dt}\begin{bmatrix} \chi_1 \\ \chi_2 \end{bmatrix} = \frac{d}{dt}\chi, \chi = \begin{bmatrix} \chi_1 \\ \chi_2 \end{bmatrix}$$

$$f = \begin{bmatrix} f_1 \\ f_2 \end{bmatrix} = \begin{bmatrix} \chi_2 \\ -\gamma_0\chi_1 - \gamma_1\chi_2 \end{bmatrix}, A = \begin{bmatrix} 0 & 1 \\ -\gamma_0 & -\gamma_1 \end{bmatrix} \tag{10.3.4}$$

The solution χ of Eq. (10.3.3) represents the state of the dynamical system as t changes. An equilibrium state χ_e is a state of the system which satisfies the equation

$$\dot{\chi}_e = 0 = f$$

A plot that may reveal the qualitative behavior of the system is the phase diagram. It is a plot of χ versus $d\chi/dt$. Phase diagrams are especially graphic when χ is a one-dimensional variable so that the phase space composed of $\{\chi, d\chi/dt\}$ is a two-dimensional space. Phase space and phase diagrams are widely used in classical and statistical mechanics.

The stability of a dynamical system can be determined by calculating what happens to the system when it is slightly perturbed from an equilibrium state. Stability calculations are relatively straightforward for linear systems, but can be very difficult or intractable for nonlinear problems. Since many dynamical models are nonlinear, approximation techniques must be used to analyze their stability.

One way to analyze the stability of a nonlinear, dynamical model is to first linearize the problem. As a first approximation, the nonlinear problem is linearized by performing a Taylor series expansion of Eq. (10.3.3) about an equilibrium point. The result is

$$\dot{u} = Ju + \zeta(u), u \equiv \chi - \chi_e \tag{10.3.5}$$

where u is the displacement of the system from its equilibrium state and $\zeta(u)$ contains terms of second-order or higher from the Taylor series expansion. The Jacobian matrix J is evaluated at the equilibrium point χ_e; thus

$$J = \begin{bmatrix} \dfrac{\partial f_1}{\partial \chi_1} & \dfrac{\partial f_1}{\partial \chi_2} \\ \dfrac{\partial f_2}{\partial \chi_1} & \dfrac{\partial f_2}{\partial \chi_2} \end{bmatrix}_{\chi_e} = \begin{bmatrix} 0 & 1 \\ -\gamma_0 & -\gamma_1 \end{bmatrix}_{\chi_e}$$

In this case, the matrices A and J are equal. Neglecting higher-order terms in Eq. (10.3.5) gives the linearized equation

$$\dot{u} = Ju \tag{10.3.6}$$

We solve Eq. (10.3.6) by trying a solution with the exponential time dependence

$$u = e^{\lambda t} g \tag{10.3.7}$$

where g is a nonzero vector and λ indicates whether or not the solution will return to equilibrium after a perturbation. Substituting Eq. (10.3.7) into Eq. (10.3.6) gives an eigenvalue problem of the form

$$(J - \lambda I)g = 0 \tag{10.3.8}$$

The eigenvalues λ are found from the characteristic equation

$$\det(J - \lambda I) = 0 \tag{10.3.9}$$

where det denotes the determinant. The following summarizes the interpretation of λ if we assume that the independent variable t is monotonically increasing:

STABILITY EIGENVALUE INTERPRETATION

Eigenvalue Condition	Interpretation
$\lambda > 0$	Diverges from equilibrium solution
$\lambda = 0$	Transition point
$\lambda < 0$	Converges to equilibrium solution

Eigenvalues from the characteristic equation for our particular case are

$$\lambda_\pm = \frac{1}{2}\left[-\gamma_1 \pm \sqrt{\gamma_1^2 - 4\gamma_0} \right] \tag{10.3.10}$$

The linearized form of Eq. (10.3.3), namely, Eq. (10.3.6), exhibits stability when the product λt is less than zero because the difference $u \to 0$ as $\lambda t \to -\infty$ in Eq. (10.3.7). If the product λt is greater than zero, the difference u diverges. This does not mean the solution of the nonlinear problem is globally divergent because of our linearization assumption. It does imply that a perturbation of the solution from its equilibrium value is locally divergent. Thus an estimate of the stability of the system is found by calculating the eigenvalues from the characteristic equation.

EXERCISE 10.7: Suppose we are given the equation $m\ddot{x} + c\dot{x} + kx = 0$ as a model of a mechanical system, where $\{m, c, k\}$ are real constants. Under what conditions are the solutions stable? *Hint*: Rearrange the equation to look like Eq. (10.3.1) and then find the eigenvalues corresponding to the characteristic equation of the stability analysis.

10.4 INTRODUCTION TO NONLINEAR DYNAMICS AND CHAOS

Nonlinear difference equations such as

$$x_{n+1} = \lambda x_n(1 - x_n) \tag{10.4.1}$$

have arisen as simple models of turbulence in fluids, fluctuation of economic prices, and the evolution of biological populations. The mathematical significance of Eq. (10.4.1) was first recognized by biologist Robert May [1976], and Eq. (10.4.1) is now known as the *May equation*. May viewed Eq. (10.4.1) as a biological model representing the evolution of a population x_n. The May equation has the quadratic form

$$x_{n+1} \equiv F(x_n) = \lambda x_n - \lambda(x_n)^2 \tag{10.4.2}$$

where $F(x_n)$ is called the logistic function. The linear term λx_n represents linear growth ($\lambda > 1$) or death ($\lambda < 1$). The nonlinear term $-\lambda(x_n)^2$ represents a nonlinear death rate that dominates when the population becomes sufficiently large. Once we specify an initial state x_0, the evolution of the system is completely determined by the relationship $x_{n+1} = F(x_n)$. The May equation is an example of a dynamical system.

Each dynamical system consists of two parts: (1) a state and (2) a dynamic. The system is characterized by the state, and the dynamic is the relationship that describes the evolution of the state.

The function $F(x_n)$ in Eq. (10.4.2) is the dynamic that maps the state x_n to the state x_{n+1}. The quantity λ in Eq. (10.4.1) is the parameter of the map F. Applying the map F for N iterations generates a sequence of states $\{x_0, x_1, x_2, \ldots, x_N\}$ known as the *orbit* of the iterative mapping as N gets very large. The behavior of the dynamical system approaches steady-state behavior as $N \to \infty$. For large N, the steady-state x_N must be bounded.

There are several useful graphs that can be made to help visualize the behavior of a dynamical system. Three are summarized as follows:

HELPFUL NONLINEAR DYNAMICAL PLOTS
Assume a discrete map $x_{i+1} = f_\lambda(x_i)$ with parameter λ.

Type	Plot	Comments
Time evolution	i vs. (x_i or x_{i+1})	Vary initial value x_0 and λ to study sensitivity to initial conditions
Return map	x_i vs. x_{i+1}	Vary initial value x_0 and λ to look for instabilities
Logistic map	x_{i+1} vs. λ	Search for order (patterns) in chaos (apparent randomness)

Example: The May Equation. May's logistic function $F(x_n)$ in Eq. (10.4.2) has a linear term and a nonlinear term. Suppose we normalize the population so that $0 \le x_n \le 1$. If the original population x_0 is much less than 1, then the nonlinear term is negligible initially and the population at the end of the first iteration is proportional to x_0. The parameter λ is the proportionality constant. The population will increase if $\lambda > 1$, and it will decrease if $\lambda < 1$. If $\lambda > 1$, the population will eventually grow until the nonlinear term dominates, at which time the population will begin to decline because the nonlinear term is negative.

The logistic map is the equation of an inverted parabola. When $\lambda < 1$, all populations will eventually become extinct, that is, x_{n+1} will go to zero. If λ is a value between 1 and 3, almost all values of x_0 approach—or are attracted to—a fixed point. If λ is larger than 3, the number of fixed points begins to increase until, for sufficiently large λ, the values of x do not converge to any fixed point and the system becomes chaotic. To see the onset of chaos, we plot return maps. Figures 10.1 and 10.2 show return maps for an initial condition $x_0 = 0.5$ and the parameter λ equals 3.5 and 3.9. There are only four fixed points when $\lambda = 3.5$. The number of fixed points increases dramatically when $\lambda = 3.9$.

EXERCISE 10.8: Another way to see the onset of chaos is to plot the number of fixed points versus the parameter λ. This plot is called the *logistic map*. Plot the logistic map for the May equation in the range $3.5 \le \lambda \le 4.0$ and initial condition $x_0 = 0.5$.

Application: Defining Chaos. Chaotic behavior of a dynamical system may be viewed graphically as two trajectories that begin with nearby initial conditions (Figure 10.3). If the trajectories are sensitive to initial conditions, so much so that they diverge at an exponential rate, then the dynamical system is exhibiting chaotic behavior. A quantitative characterization of this behavior is expressed in terms of the Lyapunov exponent [Hirsch and Smale, 1974; Lichtenberg and Lieberman, 1983; Jensen, 1987].

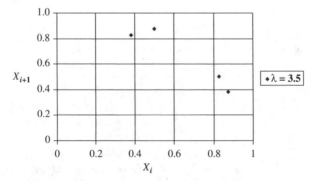

Figure 10.1   Return map of May equation at $\lambda = 3.5$.

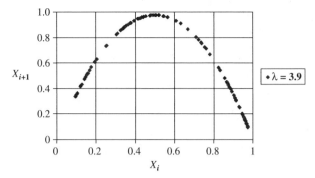

Figure 10.2 Return map of May equation at $\lambda = 3.9$.

The definition of chaos as a dynamical system with positive Lyapunov exponents [Jensen, 1987] can be motivated by deriving its one-dimensional form from the first-order differential equation

$$\frac{dx_0}{dt} = V(x_0) \tag{10.4.3}$$

A solution of Eq. (10.4.3) for a given initial condition is called a trajectory, and the set of all trajectories is the flow of Eq. (10.4.3). The flow is generated by the mapping V. Further terminology and discussion of the classification of differential equations may be found in the literature, for example, Hirsch and Smale [1974], Gluckenheimer and Holmes [1986], and Devaney [1992].

The separation Δx between two trajectories is calculated by slightly displacing the original trajectory. The difference between the original (x_0) and displaced $(x_0 + \Delta x)$ trajectories is

$$\frac{d(x_0 + \Delta x)}{dt} - \frac{dx_0}{dt} = V(x_0 + \Delta x) - V(x_0) \tag{10.4.4}$$

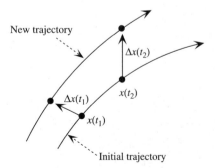

Figure 10.3 Schematic of diverging trajectories.

Substituting a first-order Taylor series expansion of $V(x_0 + \Delta x)$ in Eq. (10.4.4) gives

$$\frac{d(\Delta x)}{dt} = \left[\frac{dV(x)}{dx}\right]_{x_0} \Delta x \tag{10.4.5}$$

Equation (10.4.5) has the solution

$$\Delta x = \Delta x_0 \exp\left\{\int_0^t \left[\frac{dV}{dx}\right]_{x_0} dt'\right\} \tag{10.4.6}$$

where t' is a dummy integration variable and the separation Δx_0 is the value of Δx at $t = 0$.

A measure of the trajectory separation is obtained by calculating the magnitude of Δx. Thus the trajectory separation $d(x_0, t)$ is

$$d(x_0, t) = \left|\sqrt{(\Delta x)^2}\right| = |\Delta x| \tag{10.4.7}$$

Following Lichtenberg and Lieberman [1983], a chaotic system is a system that exhibits an exponential rate of divergence of two initially close trajectories. This concept is quantified by assuming the exponential relationship

$$d(x_0, t) = d(x_0, 0)e^{\sigma(x_0, t)t} \tag{10.4.8}$$

where $d(x_0, 0)$ is the trajectory separation at $t = 0$ and σ is a factor to be determined. A positive value of σ in Eq. (10.4.8) implies a divergence of trajectories, whereas a negative value implies convergence. Solving Eq. (10.4.8) for σ and recognizing that σt is positive for diverging trajectories lets us write

$$\sigma(x_0, t) = \frac{1}{t}\int_0^t \left[\frac{dV}{dx}\right]_{x_0} dt' \tag{10.4.9}$$

where Eqs. (10.4.6) and (10.4.7) have been used. The Lyapunov exponent σ_L is the value of $\sigma(x_0, t)$ in the limit as t goes to infinity; thus

$$\sigma_L = \lim_{t \to \infty} \sigma(x_0, t) = \lim_{t \to \infty}\left\{\frac{1}{t}\int_0^t \left[\frac{dV}{dx}\right]_{x_0} dt'\right\} \tag{10.4.10}$$

Equation (10.4.10) is an effective definition of chaos if the derivative dV/dx is known at x_0, and both the integral and limit can be evaluated. An alternative definition of the Lyapunov exponent is given by Parker and Chua [1987].

Example: Calculation of the Lyapunov Exponent. Suppose $V(x) = \lambda x$ so that Eq. (10.4.3) can be written in the explicit form

$$\frac{dx}{dt} = \lambda x$$

The derivative of V with respect to x is

$$\frac{dV}{dx} = \lambda$$

Substituting this derivative into Eq. (10.4.9) gives

$$\sigma(x_0, t) = \frac{1}{t} \int_0^t \left[\frac{dV}{dx} \right]_{x_0} dt'$$

$$= \frac{1}{t} \int_0^t \lambda dt'$$

$$= \frac{\lambda t}{t}$$

$$= \lambda$$

The Lyapunov exponent is now calculated as the limit

$$\sigma_L = \lim_{t \to \infty} \sigma(x_0, t)$$

$$= \lim_{t \to \infty}$$

$$= \lambda$$

In this case σ_L and $\sigma(x_0, t)$ are equal because $\sigma(x_0, t)$ does not depend on time. The trajectory separation Δx is given by Eq. (10.4.4) as

$$\frac{d(\Delta x)}{dt} = V(x_0 + \Delta x) - V(x_0)$$

$$= \lambda(x_0 + \Delta x) - \lambda(x_0)$$

$$= \lambda \Delta x$$

with the condition that $\Delta x = \Delta x_0$ at $t = 0$. This equation has the solution

$$\Delta x = \Delta x_0 e^{\lambda t}$$

Notice that a positive value of λ implies σ_L is positive and the trajectory separation is divergent. Similarly, a negative value of λ implies σ_L is negative and the trajectory separation is convergent.

CHAPTER 11

ODE SOLUTION TECHNIQUES

There is no guarantee that an ODE can be solved. Several different techniques exist for trying to solve ODEs, and it is often necessary to apply more than one method to ODEs that are not in a familiar form. The solution techniques presented in this chapter have been used successfully to solve a wide range of ODE problems.

11.1 HIGHER-ORDER ODE WITH CONSTANT COEFFICIENTS

Suppose we are given the nonhomogeneous ODE

$$A\ddot{x} + B\dot{x} + Cx = D\cos\omega t, D > 0, \omega > 0 \qquad (11.1.1)$$

where A, B, C, D, ω are constants, and the time derivatives are given by

$$\dot{x} = \frac{dx}{dt}, \ddot{x} = \frac{d^2x}{dt^2}$$

We seek to solve the ODE using the particular solution

$$x(t) = a\cos\omega t + b\sin\omega t \qquad (11.1.2)$$

Math Refresher for Scientists and Engineers, Third Edition By John R. Fanchi
Copyright © 2006 John Wiley & Sons, Inc.

for initial conditions

$$x(t=0) = x_0, \dot{x}(t=0) = \dot{x}_0 \tag{11.1.3}$$

The time derivatives of $x(t)$ are

$$\dot{x} = -a\omega \sin \omega t + b\omega \cos \omega t$$

$$\ddot{x} = -\omega^2 a \cos \omega t - \omega^2 b \sin \omega t$$

We substitute these expressions into the nonhomogeneous ODE to find

$$-A\omega^2(a\cos \omega t + b\sin \omega t) + B\omega(-a\sin \omega t + b\cos \omega t) + C(a\cos \omega t + b\sin \omega t)$$
$$= D\cos \omega t$$

Collecting terms in $\cos \omega t$ and $\sin \omega t$ gives

$$[-A\omega^2 a + bB\omega + aC - D]\cos \omega t$$
$$+ [-A\omega^2 b - aB\omega + bC]\sin \omega t = 0$$

This equation is satisfied if the coefficients of each trigonometric term vanish. Thus we have two equations for the two unknowns a, b:

$$a(C - A\omega^2) + bB\omega = D$$
$$a(-B\omega) + b(C - A\omega^2) = 0$$

Solving the latter equation for b yields

$$b = \frac{B\omega}{C - A\omega^2}a$$

Substituting the expression for b into the equation containing D gives

$$a(C - A\omega^2) + \frac{B^2\omega^2}{C - A\omega^2}a = D$$

We simplify these equations to obtain

$$a = \frac{D(C - A\omega^2)}{(C - A\omega^2)^2 + B^2\omega^2} \tag{11.1.4}$$

and

$$b = \frac{DB\omega}{(C - A\omega^2)^2 + B^2\omega^2} \tag{11.1.5}$$

At $t = 0$, the initial conditions are

$$x(0) = a = x_0, \dot{x}(0) = b\omega = \dot{x}_0 \tag{11.1.6}$$

Example: Application to a Mechanical System, Forced Oscillations. Define the following:

$\quad A = m = $ mass of the body

$\quad B = c = $ damping constant

$\quad C = k = $ spring modulus

$\quad D = F_0 = $ applied external force at $t = 0$ that corresponds to maximum
$\qquad\qquad$ driving force

$\quad x(t) = $ displacement of spring from equilibrium

Using this notation, the nonhomogeneous ODE for the mechanical system shown in Figure 11.1 is

$$m\ddot{x} + c\dot{x} + kx = F_0 \cos \omega t$$

From the preceding discussion we have a particular solution

$$x(t) = a \cos \omega t + b \sin \omega t$$

where

$$a = F_0 \frac{(k - m\omega^2)}{(k - m\omega^2)^2 + c^2\omega^2}$$

$$b = F_0 \frac{c\omega}{(k - m\omega^2)^2 + c^2\omega^2}$$

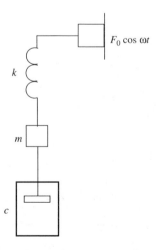

Figure 11.1   Mass on a damped spring.

If we define a natural mechanical frequency

$$\omega_0 = \sqrt{k/m} > 0$$

we obtain

$$a = F_0 \frac{m(\omega_0^2 - \omega^0)}{m^2(\omega_0^2 - \omega^2)^2 + \omega^2 c^2}$$

$$b = F_0 \frac{\omega c}{m^2(\omega_0^2 - \omega^2)^2 + \omega^2 c^2}$$

Resonance occurs when $\omega_0 = \omega$. In an undamped system (when $c = 0$), resonance can result in large oscillations because $a \to 0$ and $b \to \infty$. In the case of mechanical oscillations, physical values for the initial conditions are

$$x_0 = a, \dot{x}_0 = b\omega$$

EXERCISE 11.1: Application to an Electrical System—*RLC* Circuit. Find a particular solution of the nonhomogeneous ODE

$$L\ddot{I} + R\dot{I} + \frac{I}{\hat{C}} = E_0\omega \cos \omega t$$

where $L =$ inductance of inductor, $R =$ resistance of resistor, $\hat{C} =$ capacitance of capacitor, $E_0\omega =$ maximum value of derivative of electromotive force ($E_0 =$ voltage at $t = 0$), and $I(t) =$ current (see Figure 11.2).

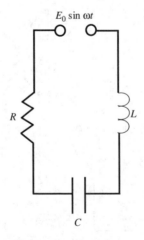

Figure 11.2 *RLC* circuit.

11.2 VARIATION OF PARAMETERS

Consider the second-order ODE

$$y'' + a_1(x)y' + a_0(x)y = h(x) \tag{11.2.1}$$

where $h(x)$ is a known function for the nonhomogeneous equation and the prime denotes differentiation with respect to x. The double prime denotes the second-order derivative with respect to x. Equation (11.2.1) has the homogeneous ($h(x) = 0$) solution

$$y_h = c_1 y_1(x) + c_2 y_2(x)$$

where c_1, c_2 are constant parameters, and y_1, y_2 are solutions of the homogeneous equation. To solve the nonhomogeneous equation, we seek a particular solution y_p for the boundary conditions

$$y_p(x_0) = 0, y'_p(x_0) = 0 \tag{11.2.2}$$

The method we use follows Section 3.4 of Kreider et al. [1968].
    We try a solution of the form

$$y_p = c_1(x)y_1(x) + c_2(x)y_2(x) \tag{11.2.3}$$

where parameters of the homogeneous solution are now allowed to depend on x. Substitute Eq. (11.2.3) into Eq. (11.2.1) to get

$$c_1\left(y_1'' + a_1 y_1' + a_0 y_1\right) + c_2\left(y_2'' + a_1 y_2' + a_0 y_2\right) + \left(c_1'' y_1 + c_2'' y_2\right)' \\ + a_1\left(c_1' y_1 + c_2' y_2\right) + \left(c_1' y_1' + c_2' y_2'\right) = h$$

The terms multiplying c_1 and c_2 must equal 0 since y_1 and y_2 solve the homogeneous equation with $h(x) = 0$. Simplifying the above equation gives

$$c_1'' y_1 + c_1' + y_1' + c_2'' y_2 + c_2' y_2' + a_1\left(c_1' y_1 + c_2' y_2\right) + \left(c_1' y_1' + c_2' y_2'\right) = h$$

Upon rearrangement we find

$$\frac{d}{dx}\left(c_1' y_1 + c_2' y_2\right) + a_1\left(c_1' y_1 + c_2' y_2\right) + \left(c_1' y_1' + c_2' y_2'\right) = h \tag{11.2.4}$$

Equation (11.2.4) is solved if the following conditions are satisfied by the arbitrary parameters c_1, c_2:

$$c_1' y_1 + c_2' y_2 = 0 \tag{11.2.5a}$$
$$c_1' y_1' + c_2' y_2' = h \tag{11.2.5b}$$

Rearranging Eq. (11.2.5a) gives

$$c_1' = -\frac{c_2' y_2}{y_1} \tag{11.2.6}$$

which may be used in Eq. (11.2.5b) to yield

$$-c_2' \frac{y_2}{y_1} y_1' + c_2' y_2' = c_2'\left[y_2' - \frac{y_2}{y_1} y_{i'}'\right] = h$$

Solving for c_2' and simplifying lets us write

$$c_2' = \frac{h y_1}{y_1 y_2' - y_2 y_1'} \tag{11.2.7}$$

Using Eq. (11.2.7) in Eq. (11.2.6) gives

$$c_1' = -\frac{y_2}{y_1} c_2' = \frac{-h y_2}{y_1 y_2' - y_2 y_1'} \tag{11.2.8}$$

Note that the denominator may be written as the determinant

$$y_1 y_2' - y_2 y_1' = \begin{vmatrix} y_1 & y_2 \\ y_1' & y_2' \end{vmatrix} \equiv W(y_1, y_2) \tag{11.2.9}$$

The determinant is called the *Wronskian* and is denoted by $W(y_1, y_2)$. The Wronskian notation lets us simplify Eqs. (11.2.7) and (11.2.8); thus

$$c_1' = \frac{dc_1}{dx} = -\frac{h y_2}{W(y_1, y_2)}, c_2' = \frac{dc_2}{dx} = \frac{h y_1}{W(y_1, y_2)} \tag{11.2.10}$$

If we integrate c_1' and c_2' over a dummy variable t from a fixed reference point x_0 to a variable upper limit x, the particular solution becomes

$$y_p = c_1 y_1 + c_2 y_2$$

or

$$y_p = \left\{\int_{x_0}^{x}\left[-\frac{h(t) y_2(t)}{W[y_1(t), y_2(t)]}\right] dt\right\} y_1(x)$$

$$+ \left\{\int_{x_0}^{x} \frac{h(t) y_1(t)}{W[y_1(t), y_2(t)]} dt\right\} y_2(x)$$

Combining terms gives

$$y_p(x) = \int_{x_0}^{x}\left\{\frac{y_1(t) y_2(x) - y_1(x) y_2(t)}{W[y_1(t), y_2(t)]}\right\} h(t)\, dt \tag{11.2.11}$$

since $y_1(x)$ and $y_2(x)$ are independent of the dummy integration variable t. Equation (11.2.11) can be written as

$$y_p(x) = \int_{x_0}^{x} K(x,t)h(t)\,dt \qquad (11.2.12)$$

where

$$K(x,t) = \frac{y_1(t)y_2(x) - y_1(x)y_2(t)}{W[y_1(t), y_2(t)]} \qquad (11.2.13)$$

The function $K(x,t)$ is called the *Green's function* for the nonhomogeneous equation $Ly = h(x)$ with the differential operator L given by

$$L = \frac{d^2}{dx^2} + a_1(x)\frac{d}{dx} + a_0(x) \qquad (11.2.14)$$

Inspecting Eqs. (11.2.9) and (11.2.13) shows that Green's function depends only on the solutions y_1, y_2 of the homogeneous equation $Ly = 0$. Green's function $K(x,t)$ does not depend on the function $h(x)$. Once $K(x,t)$ has been calculated, Eq. (11.2.12) gives a particular solution in terms of the function $h(x)$. In summary, the general solution of $Ly = h(x)$ is

$$y(x) = y_p(x) + y_h(x) \qquad (11.2.15)$$

or

$$y(x) = \int_{x_0}^{x} K(x,t)h(t)\,dt + y_h(x) \qquad (11.2.16)$$

EXERCISE 11.2: Find the general solution of the one-dimensional Schrödinger equation

$$-\frac{\hbar^2}{2m}\frac{d^2}{dx^2}\psi(x) - E\psi(x) = V(x)$$

for the wavefunction $\psi(x)$. The two boundary conditions are $\psi(x_0) = 0, \psi'(x_0) = 0$.

11.3 CAUCHY EQUATION

Another ODE solution technique is illustrated by solving an equation of the form

$$x^2 y'' + axy' + by = 0 \qquad (11.3.1)$$

where a, b are constants. This equation is called the *Cauchy* (or *Euler*) equation. It is a second-order ODE and should therefore have two solutions. To find them, we begin by assuming

$$y = x^m \tag{11.3.2}$$

and substitute y into the Cauchy equation to get

$$x^2 m(m-1) x^{m-2} + axmx^{m-1} + bx^m = 0$$

Combining factors of x lets us simplify the equation to the form

$$m(m-1)x^m + amx^m + bx^m = 0$$

Factoring x^m gives

$$[m(m-1) + am + b] x^m = 0 \tag{11.3.3}$$

For arbitrary values of x, the coefficient in brackets $[\cdots]$ must vanish, and we obtain the indicial equation

$$[m(m-1) + am + b] = 0$$

or

$$m^2 + (a-1)m + b = 0$$

The indicial equation is a quadratic equation with the roots

$$m_\pm = -\frac{(a-1)}{2} \pm \frac{\sqrt{(a-1)^2 - 4b}}{2} \tag{11.3.4}$$

Hence the solution of Eq. (11.3.1) is

$$y = c_1 x^{m_+} + c_2 x^{m_-} \tag{11.3.5}$$

for

$$m_+ \neq m_-$$

For the special case of a double root ($m_+ = m_-$), only one solution is found and another procedure such as variation of parameters must be used to obtain the other solution.

11.4 SERIES METHODS

Consider a linear, second-order ODE of the form

$$A(x)y'' + B(x)y' + C(x)y = 0 \tag{11.4.1}$$

where the prime denotes differentiation with respect to x. The functions $\{A(x), B(x), C(x)\}$ are analytic functions; that is, each function exists and is differentiable at all points in a domain D (see Chapter 7, Section 7.3 for more discussion of analyticity). Equation (11.4.1) is a homogeneous ODE with the general solution

$$y(x) = \alpha_1 y_1(x) + \alpha_2 y_2(x) \tag{11.4.2}$$

where $\{\alpha_1, \alpha_2\}$ are constants, and $\{y_1, y_2\}$ are two linearly independent solutions of Eq. (11.4.1). Two powerful methods for finding solutions of Eq. (11.4.1) are discussed in this section: (1) the power series method and (2) the Frobenius series method. Both methods depend on series representations of the solution $y(x)$.

Power Series Method

The power series method is usually presented as a technique for solving Eq. (11.4.1) in the more traditional form

$$y'' + P(x)y' + Q(x)y = 0 \tag{11.4.3}$$

Equation (11.4.3) was obtained by dividing Eq. (11.4.1) by $A(x)$ so that

$$P(x) = \frac{B(x)}{A(x)}, \, Q(x) = \frac{C(x)}{A(x)} \tag{11.4.4}$$

If the functions $P(x)$ and $Q(x)$ are both analytic at $x = 0$, then $x = 0$ is said to be an ordinary point. In this case, Eq. (11.4.2) has two linearly independent power series solutions of the form

$$y(x) = \sum_{n=0}^{\infty} c_n x^n \tag{11.4.5}$$

with expansion coefficients $\{c_n\}$. An example illustrating the procedure for calculating the expansion coefficients $\{c_n\}$ follows.

Example: Legendre's equation is

$$(1 - x^2)y'' - 2xy' + my = 0$$

where m is a real number and the prime denotes differentiation with respect to x. Legendre's equation is solved using the power series expansion

$$y(x) = \sum_{k=0}^{\infty} c_k x^k$$

The derivatives of y are

$$y' = \sum_{k=1}^{\infty} k c_k x^{k-1}$$

and

$$y'' = \sum_{k=2}^{\infty} k(k-1) c_k x^{k-2}$$

Substituting the derivatives into Legendre's equation gives

$$(1-x^2) \sum_{k=2}^{\infty} k(k-1) c_k x^{k-2} - 2x \sum_{k=1}^{\infty} k c_k x^{k-1} + m \sum_{k=0}^{\infty} c_k x^k = 0$$

Multiplying the factors in the first and second summations yields

$$\sum_{k=2}^{\infty} k(k-1) c_k x^{k-2} - \sum_{k=2}^{\infty} k(k-1) c_k x^k$$

$$-2 \sum_{k=1}^{\infty} k c_k x^k + m \sum_{k=0}^{\infty} c_k x^k = 0$$

Writing out the terms and collecting like powers of x lets us write

$$m c_0 + m c_1 x + m c_2 x^2 + m c_3 x^3 + \cdots + m c_r x^r + \cdots$$

$$+ 2 c_2 + 3 \cdot 2 c_3 x + 4 \cdot 3 c_4 x^2 + \cdots + (r+2)(r+1) c_{r+2} x^r + \cdots$$

$$- 2 c_2 x^2 - 3 \cdot 2 c_3 x^3 - \cdots - r(r-1) c_r x^r - \cdots$$

$$- 2 c_1 x - 2 \cdot 2 c_2 x^2 - 2 \cdot 3 c_3 x^3 - \cdots - 2 r c_r x^r - \cdots = 0$$

This equation is satisfied for all values of x when the coefficients of each power of x vanish; thus

$$c_2 = -\frac{m}{2} c_0$$

$$c_3 = -\frac{m-2}{3 \cdot 2} c_1$$

$$c_4 = -\frac{m - 2 \cdot 2 - 2}{4 \cdot 3} c_2$$

$$c_5 = -\frac{m - 2 \cdot 3 - 3 \cdot 2}{5 \cdot 4} c_3$$

etc.

In general, the expansion coefficients satisfy the relation

$$c_{r+2} = -\frac{m - 2r - r(r-1)}{(r+2)(r+1)} c_r$$

This relationship between expansion coefficients is an example of a recurrence relation.

The solution of Legendre's equation can be expressed as the sum of two polynomials using the recurrence relation. First write the power series solution in the form

$$y(x) = \sum_{k=0}^{\infty} c_k x^k$$

$$= c_0 + c_2 x^2 + c_4 x^4 + \cdots$$

$$+ c_1 x + c_3 x^3 + c_5 x^5 + \cdots$$

The coefficients $\{c_k\}$ with $k > 1$ are now written in terms of c_0 and c_1 using the recurrence relations to give

$$y(x) = c_0 \left[1 - \frac{m}{2} x^2 + \frac{m}{2} \left(\frac{m-4-2}{4 \cdot 3} \right) x^4 - \cdots \right]$$

$$+ c_1 \left[x - \frac{m-2}{3 \cdot 2} x^2 + \frac{m-2}{3 \cdot 2} \left(\frac{m - 2 \cdot 3 - 3 \cdot 2}{5 \cdot 4} \right) x^5 - \cdots \right]$$

or

$$y(x) = c_0 y_1(x) + c_1 y_2(x)$$

The polynomials $y_1(x)$ and $y_2(x)$ are linearly independent power series. The constants c_0 and c_1 can be determined when two boundary conditions are specified for Legendre's second-order ODE.

Frobenius Series Method

The power series solution method just described is applicable to Eq. (11.4.3) when $P(x)$ and $Q(x)$ are analytic at $x = 0$. If $P(x)$ or $Q(x)$ is not analytic at $x = 0$, then $x = 0$ is called a singular point and Eq. (11.4.3) is rewritten in the form

$$y'' + \frac{p(x)}{x} y' + \frac{q(x)}{x^2} y = 0 \tag{11.4.6}$$

where

$$p(x) = xP(x), \quad q(x) = x^2 Q(x) \tag{11.4.7}$$

If $p(x)$ and $q(x)$ are both analytic at $x = 0$, then $x = 0$ is said to be a regular singular point. In this case, a solution of Eq. (11.4.6) is the Frobenius series

$$y(x) = x^r \sum_{n=0}^{\infty} a_n x^n \tag{11.4.8}$$

where $\{a_n\}$ and r are constants. It is assumed that the variable x is restricted to the interval $0 < x < \rho$, where ρ is the radius of convergence of the series (see Chapter 2, Section 2.5). The restriction on x positive values can be relaxed if we replace x^r by $|x|^r$ in Eq. (11.4.8). In many applications, however, the ODE in Eq. (11.4.6) arises from physical models expressed in radial coordinates, and the variable x in Eq. (11.4.6) then represents the non negative radial coordinate. The constant r in Eq. (11.4.8) is found by substituting Eq. (11.4.8) into Eq. (11.4.6) or its equivalent

$$x^2y'' + xp(x)y' + q(x)y = 0 \tag{11.4.9}$$

The result is the indicial equation

$$r(r-1) + p_0 r + q_0 = 0 \tag{11.4.10}$$

with the notation $p_0 = p(0)$ and $q_0 = q(0)$. An indicial equation was encountered in Section 11.3 as part of the solution of the Cauchy equation. The two roots of the indicial equation are

$$r = \tfrac{1}{2}\left[1 - p_0 \pm \sqrt{(1 - p_0)^2 - 4q_0}\right] \tag{11.4.11}$$

Notice that the roots of the indicial equation are not necessarily integers.

At least one Frobenius series solution of Eq. (11.4.9) exists when the roots of Eq. (11.4.10) are real. In this case, one solution of Eq. (11.4.9) is

$$y_1(x) = x^{r_1} \sum_{n=0}^{\infty} a_n x^n \tag{11.4.12}$$

where $r_1 > r_2$ and $a_0 \neq 0$. A second linearly independent solution of Eq. (11.4.9) exists if $r_1 - r_2 \neq 0$ and $r_1 - r_2$ is not a positive integer. The second solution is then

$$y_2(x) = x^{r_2} \sum_{n=0}^{\infty} b_n x^n \tag{11.4.13}$$

with $b_0 \neq 0$.

The coefficients in Eqs. (11.4.12) and (11.4.13) are found by substituting the appropriate Frobenius series into Eq. (11.4.9) and expanding the resulting set of equations in equal powers of x until a recurrence relation is found. The Frobenius series solution method is illustrated in the following example and Exercise 11.3. Additional discussion of series solution methods can be found in many sources, for example, Lomen and Mark [1988], Edwards and Penney [1989], and Kreyszig [1999].

Example: Solve $y'' + y = 0$ using a power series solution of the form

$$y(x) = \sum_{m=0}^{\infty} c_m x^{m+r} \tag{11.4.14}$$

The derivatives of y with respect to x are

$$y' = \sum_{m=0}^{\infty} c_m (m+r) x^{m+r-1},$$

$$y'' = \sum_{m=0}^{\infty} c_m (m+r)(m+r-1) x^{m+r-2}$$

Substituting these power series expansions of y, y'' into the ODE gives

$$y'' + y = \sum_{m=0}^{\infty} c_m (m+r)(m+r-1) x^{m+r-2} + \sum_{m=0}^{\infty} c_m x^{m+r} = 0 \tag{11.4.15}$$

or

$$c_0 r(r-1) x^{r-2} + (r+1)r c_1 x^{r-1} + c_2 (r+2)(r+1) x^r$$
$$+ c_3 (r+3)(r+2) x^{r+1} + \cdots \tag{11.4.16}$$
$$+ c_0 x^r + c_1 x^{r+1} + c_2 x^{r+2} + \cdots = 0$$

where we have expanded the sums of Eq. (11.4.15). Equating like powers of x in Eq. (11.4.16) yields the following set of equations:

$$r(r-1) c_0 = 0$$
$$(r+1) r c_1 = 0$$
$$c_0 + c_2 (r+2)(r+1) = 0$$
$$c_1 + c_3 (r+3)(r+2) = 0$$
$$\vdots$$

The first two equations imply $r = 0$ for nonzero values of c_0 and c_1. The latter equations have the form

$$c_i + c_{i+2} (r+i+2)(r+i+1) = 0 \tag{11.4.17}$$

Solving Eq. (11.4.17) for c_{i+2} in terms of c_i leads to the recurrence relation

$$c_{i+2} = \frac{-c_i}{(r+i+2)(r+i+1)}$$

Recognizing that $r = 0$ gives

$$c_{i+2} = \frac{-c_i}{(i+2)(i+1)} \tag{11.4.18}$$

Each coefficient c_i for $i > 1$ can be expressed in terms of either c_0 or c_i; thus

$$c_2 = \frac{-c_0}{2 \cdot 1}, c_3 = \frac{-c_1}{3 \cdot 2},$$

$$c_4 = \frac{-c_2}{4 \cdot 3} = \frac{c_0}{4!}, c_5 = \frac{-c_3}{5 \cdot 4} = \frac{c_1}{5!}, \cdots \tag{11.4.19}$$

The constants c_0, c_1 are determined by boundary conditions. Substituting the expressions in Eq. (11.4.19) into the power series expansion of y gives

$$y(x) = c_0 + c_1 x - \frac{c_0}{2!}x^2 - \frac{c_1}{3!}x^3 + \frac{c_0}{4!}x^4 + \frac{c_1}{5!}x^5 + \cdots$$

or

$$y(x) = c_0 \left[1 - \frac{x^2}{2!} + \frac{x^4}{4!} - + \cdots \right] + c_1 \left[x - \frac{x^3}{3!} + \frac{x^5}{5!} - + \cdots \right] \tag{11.4.20}$$

$$= c_0 \cos x + c_1 \sin x$$

where we have used the Maclaurin series expansions for $\sin x$ and $\cos x$ (see Chapter 2, Section 2.5).

In this example, the power series method yielded a solution that we might have guessed by inspection of the ODE. Although the power series method was not needed to solve this exercise, doing so demonstrated its utility for problems where a trial solution is not so obvious.

EXERCISE 11.3: Use the Frobenius method to solve the equation

$$\left[\frac{d^2}{dr^2} + \frac{2}{r}\frac{d}{dr} - \frac{\varepsilon}{r^2} + \omega^2 \right] R(r) = 0$$

for $R(r)$ when ε and ω are independent of the variable r, and $r > 0$.

11.5 LAPLACE TRANSFORM METHOD

The Laplace transforms of derivatives presented in Section 9.6 can be used to solve differential equations by performing the following procedure:

1. Take the Laplace transform of the differential equation.
2. Solve the resulting algebraic equation for the Laplace transform.
3. Invert the Laplace transformation.
4. Verify the results.

This procedure is demonstrated by the following example.

Example: Solve the ODE $y''(x) + y(x) = 0$ with initial conditions $y(0) = a$, $y'(0) = b$ using Laplace transforms.

Step 1: Write the Laplace transform of $y(x)$ as $Y(s) = \mathcal{L}\{y(x)\}$ and take the Laplace transform of $y''(x) + y(x) = 0$ to obtain

$$\mathcal{L}\{y''(x)\} + \mathcal{L}\{y(x)\} = 0$$

or

$$s^2 Y(s) - sy(0) - y'(0) + Y(s) = 0$$

Step 2: Solve for $Y(s)$ and use the initial conditions to get

$$Y(s) = \frac{sa + b}{s^2 + 1} = \frac{sa}{s^2 + 1} + \frac{b}{s^2 + 1}$$

Step 3: To find $y(x)$, we invert $Y(s)$:

$$y(x) = \mathcal{L}^{-1}\{Y(s)\} = a \cos x + b \sin x$$

Step 4: This solution is verified by substitution into the ODE:

$$y''(x) = \frac{d^2}{dx^2}(a \cos x + b \sin x) = -a \cos x - b \sin x = -y(x)$$

EXERCISE 11.4: Use the Laplace transform to solve the ODE

$$y'' + (a + b)y' + aby = 0$$

where the prime denotes differentiation with respect to x, and the boundary conditions are

$$y(0) = \alpha, y'(0) = \beta$$

The quantities a, b, α, β are constants.

CHAPTER 12

PARTIAL DIFFERENTIAL EQUATIONS

Solutions of ODEs depend on only one independent variable. Suppose a function Ψ of two or more independent variables $\{x, y, \ldots\}$ satisfies an equation of the form

$$F(x, y, \ldots, \Psi, \Psi_x, \Psi_y, \ldots, \Psi_{xx}, \Psi_{yy}, \ldots) = 0$$

where we are using the notation

$$\Psi_x = \frac{\partial \Psi}{\partial x}, \Psi_y = \frac{\partial \Psi}{\partial y}, \ldots, \Psi_{xx} = \frac{\partial^2 \Psi}{\partial x^2}, \Psi_{xy} = \frac{\partial^2 \Psi}{\partial x \partial y}, \Psi_{yy} = \frac{\partial^2 \Psi}{\partial y^2}, \ldots$$

to represent partial derivatives. We say the function $\Psi(x, y, \ldots)$ is a solution of a partial differential equation (PDE). The order of the PDE is the order of the highest derivative that appears in the equation $F(\ldots) = 0$. A PDE is linear if (1) it is first-order in the unknown functions and their derivatives; and (2) the coefficients of the derivatives are either constant or depend on the independent variables $\{x, y, \ldots\}$; that is, the PDE does not contain a product of derivatives. Several PDEs involving one or more unknown functions constitute a system of PDEs.

This chapter focuses on PDEs that are first or second order only. These PDEs occur frequently in practice. Before presenting a scheme for classifying different types of second-order PDEs, we first summarize the most common types of boundary conditions. Methods for solving PDEs are then described.

Math Refresher for Scientists and Engineers, Third Edition By John R. Fanchi
Copyright © 2006 John Wiley & Sons, Inc.

12.1 BOUNDARY CONDITIONS

Boundary conditions for commonly occurring second-order PDEs may be written in the form

$$\alpha(x, y)\, \psi(x, y) + \beta(x, y)\, \psi_n(x, y) = \gamma(x, y)$$

where $\psi(x, y)$ is the unknown function and its derivative normal to a boundary is

$$\psi_n = \frac{\partial \psi}{\partial n}$$

The functions α, β, γ are specified functions of x, y. A classification of boundary conditions is summarized below. The functions α, β, γ, and $\psi(x, y)$ and all applicable derivatives are defined in a domain R bounded by a surface S.

Name	Form	Comment
Dirichlet (first kind)	$\beta = 0$	ψ specified at S
Neumann (second kind)	$\alpha = 0$	ψ_n specified at S
Cauchy	Two equations: $\alpha = 0$ in one and $\beta = 0$ in the other	
Robbins (third kind)	α and $\beta \neq 0$	

12.2 PDE CLASSIFICATION SCHEME

Consider the second-order PDE

$$A\psi_{xx} + 2B\psi_{xy} + C\psi_{yy} = G$$

The functions A, B, C, G are known functions of x, y and the first-order partial derivatives ψ_x, ψ_y in a domain R bounded by a surface S. Many phenomena in the real world are modeled successfully using PDEs of this type. For example, see sources like Arfken and Weber [2001], John [1978], Kreyszig [1999], and Lomen and Mark [1988]. The mathematical properties and solutions of PDEs depend on their boundary conditions and the relationship between the functions A, B, C. Three examples are as follows:

PDE Type	Criterion	Example	
Hyperbolic	$B^2 > AC$	$\psi_{xx} - \psi_{tt} = 0$	Wave equation
Elliptic	$B^2 < AC$	$\psi_{xx} + \psi_{yy} = 0$	Laplace equation
Parabolic	$B^2 = AC$	$\psi_{xx} = \psi_t$	Heat equation

The boundary conditions associated with the examples above are presented as follows:

PDE Type	Criterion	Boundary Condition
Hyperbolic	$B^2 > AC$	Cauchy
Elliptic	$B^2 < AC$	Dirichlet or Neumann
Parabolic	$B^2 = AC$	Dirichlet or Neumann

12.3 ANALYTICAL SOLUTION TECHNIQUES

Separation of Variables

One of the most successful techniques for solving a PDE begins with the separation of variables to obtain a system of decoupled ODEs. We illustrate this method by solving the wave equation

$$\frac{\partial^2 \psi}{\partial x^2} - \frac{\partial^2 \psi}{\partial t^2} = 0$$

for the function $\psi(x, t)$.

The wave equation is a hyperbolic PDE. We wish to solve it by using the separation of variables technique. This means that we let the solution $\psi(x, t)$ be a product of functions $X(x)$ and $T(t)$ such that

$$\psi(x, t) = X(x)T(t)$$

Substituting this form of $\psi(x, t)$ into the wave equation gives

$$\frac{\partial^2 X(x)T(t)}{\partial x^2} - \frac{\partial^2 X(x)T(t)}{\partial t^2} = 0$$

Factoring terms and rearranging lets us write

$$T(t)\frac{\partial^2 X(x)}{\partial x^2} = X(x)\frac{\partial^2 T(t)}{\partial t^2}$$

We now divide by $X(x)T(t)$ to find

$$\frac{T(t)}{X(x)T(t)}\frac{\partial^2 X(x)}{\partial x^2} = \frac{X(x)}{X(x)T(t)}\frac{\partial^2 T(t)}{\partial t^2}$$

or

$$\frac{1}{X(x)}\frac{d^2 X(x)}{dx^2} = \frac{1}{T(t)}\frac{d^2 T(t)}{dt^2}$$

Notice that partial derivatives have been replaced by ordinary derivatives because X depends only on x and T depends only on t. Each side of the equation depends on arbitrary independent variables; hence they must equal a constant to assure equivalence. Calling the separation constant K, we have the following two equations:

$$\frac{1}{X(x)}\frac{d^2X(x)}{dx^2} = K, \quad \frac{1}{T(T)}\frac{d^2T(t)}{dt^2} = K$$

These equations are rearranged to give

$$\frac{d^2X(x)}{dx^2} = K\,X(x), \quad \frac{d^2T(t)}{dt^2} = K\,T(t)$$

which are two ODEs for the functions $X(x)$ and $T(t)$. Each equation depends on only one independent variable. If the ODEs can be solved, including the effects of boundary conditions, the product of the resulting solutions yields the solution $\psi(x, t)$ of the original PDE.

Example: A particular solution of Laplace's equation

$$\nabla^2\Phi = \left(\frac{\partial^2}{\partial x^2} + \frac{\partial^2}{\partial y^2} + \frac{\partial^2}{\partial z^2}\right)\Phi(x, y, z) = 0$$

can be found using a separable solution of the form

$$\Phi(x, y, z) = X(x)Y(y)Z(z) \tag{12.3.1}$$

Substituting Eq. (12.3.1) into $\nabla^2\Phi = 0$ and dividing by XYZ yields

$$\frac{1}{X}\frac{d^2X}{dx^2} + \frac{1}{Y}\frac{d^2Y}{dy^2} + \frac{1}{Z}\frac{d^2Z}{dz^2} = 0 \tag{12.3.2}$$

The variables may now be separated to yield three ordinary differential equations:

$$\frac{d^2X}{dx^2} = -\alpha^2 X$$

$$\frac{d^2Y}{dy^2} = -\beta^2 Y \tag{12.3.3}$$

$$\frac{d^2Z}{dz^2} = \delta^2 Z$$

The separation constants α, β, δ must satisfy the relationship

$$\delta^2 = \alpha^2 + \beta^2 \tag{12.3.4}$$

so that the ODEs in Eq. (12.3.3) satisfy Eq. (12.3.2).

The ODEs in Eq. (12.3.3) are eigenvalue equations and the separation constants $-\alpha^2, -\beta^2, \delta^2$ are the eigenvalues of the differential operators $d^2/dx^2, d^2/dy^2$,

d^2/dz^2, respectively. This is readily seen by writing the ODE for $X(x)$ as the eigen-value equation

$$L_x X = \lambda_x X \tag{12.3.5}$$

where L_x is the differential operator

$$L_x = \frac{d^2}{dx^2}$$

and the eigenvalue λ_x is expressed in terms of a constant α:

$$\lambda_x = -\alpha^2$$

The eigenvalues $\{\alpha, \beta, \delta\}$ are related by Eq. (12.3.4).

The solutions of the ODEs in Eq. (12.3.3) can be written in terms of exponential functions as

$$X(x) = a_1 e^{i\alpha x} + a_2 e^{-i\alpha x}$$
$$Y(y) = b_1 e^{i\beta y} + b_2 e^{-i\beta y} \tag{12.3.6}$$
$$Z(z) = c_1 e^{\delta z} + c_2 e^{-\delta z}$$

Alternatively, the solutions can be expressed in terms of trigonometric functions as

$$X(x) = a_1' \sin \alpha x + a_2' \cos \alpha x$$
$$Y(y) = b_1' \sin \beta y + b_2' \cos \beta y \tag{12.3.7}$$
$$Z(z) = c_1' \sinh \delta z + c_2' \cosh \delta z$$

A particular solution is found by forming the product in Eq. (12.3.1). The constants in Eq. (12.3.6) or (12.3.7) depend on the choice of boundary conditions.

EXERCISE 12.1: Assume the boundary conditions of Laplace's equation are

$$\Phi(0, y, z) = \Phi(x_B, y, z) = 0$$
$$\Phi(x, 0, z) = \Phi(x, y_B, z) = 0$$
$$\Phi(x, y, 0) = 0$$
$$\Phi(x, y, z_B) = \Phi_B(x, y)$$

in the intervals $0 \le x \le x_B, 0 \le y \le y_B$, and $0 \le z \le z_B$. Find the solution of Laplace's equation using these boundary conditions and the trigonometric solutions in Eq. (12.3.7).

EXERCISE 12.2: Separation of Variables. Find the solution of

$$\frac{\partial^2 \psi}{\partial x^2} = \frac{\partial \psi}{\partial t}$$

by assuming

$$\psi(x, t) = X(x)T(t)$$

Variable Transformation

The solution of a differential equation often depends on insight into the type of problem being solved. For example, if the differential equation is a model of a wave-like phenomenon in one space dimension, then a trial solution of the form $f(x - vt)$ is a good starting point, where v is the velocity of the traveling wave. If we set $z = x - vt$, it may be possible to reduce the governing PDE to an ordinary differential equation. This variable transformation technique is illustrated here for a convection–diffusion (CD) equation of the form

$$D\frac{\partial^2 u}{\partial x^2} - v\frac{\partial u}{\partial x} = \frac{\partial u}{\partial t} \tag{12.3.8}$$

The terms D and v are often identified as diffusion and velocity, and the unknown function $u(x, t)$ of the two independent variables x, t is concentration. At this point, we have not restricted the functional form of D and v; that is, D and v can be functions of any of the variables $\{x, t, u\}$. This is important to note because the variable transformation technique used here can generate analytical solutions of a select class of nonlinear PDEs, as we now show.

Let us make the variable transformation $u(x, t) = g(z)$, where the new variable $z = kx - \omega t$ represents a traveling wave with wavenumber k and frequency ω. The partial derivatives in Eq. (12.3.8) become

$$\frac{\partial g}{\partial t} = \frac{\partial z}{\partial t}\frac{\partial g}{\partial z} = -\omega\frac{\partial g}{\partial z}, \quad \frac{\partial g}{\partial x} = \frac{\partial z}{\partial x}\frac{\partial g}{\partial z} = k\frac{\partial g}{\partial z}$$

$$\frac{\partial^2 g}{\partial x^2} = \frac{\partial}{\partial x}\frac{\partial g}{\partial x} = \frac{\partial}{\partial x}k\frac{\partial g}{\partial z} = k\frac{\partial}{\partial z}\frac{\partial g}{\partial x} = k^2\frac{\partial^2 g}{\partial z^2}$$

Substituting these derivatives into Eq. (12.3.8) gives

$$Dk^2\frac{\partial^2 g}{\partial z^2} - vk\frac{\partial g}{\partial z} = -\omega\frac{\partial g}{\partial z} \tag{12.3.9}$$

Let us now require that D and v depend only on z and g so that we can write Eq. (12.3.9) as the ODE

$$\frac{d^2 g}{dz^2} - \frac{v}{Dk}\frac{dg}{dz} + \frac{\omega}{Dk^2}\frac{dg}{dz} = 0 \tag{12.3.10}$$

Equation (12.3.10) can be reduced to a quadrature if the two terms with first-order derivatives can be written as total differentials. We therefore restrict the form of

D and v so that D is a constant and v has the functional form $v = a + bg^n$ so that Eq. (12.3.10) becomes

$$\frac{d}{dz}\left[\frac{dg}{dz} - \left(ag + \frac{bg^{n+1}}{n+1}\right)\frac{1}{Dk} + \frac{\omega g}{Dk^2}\right] = 0 \qquad (12.3.11)$$

Equation (12.3.11) is satisfied if the bracketed term [·] is a constant c; thus

$$\frac{dg}{dz} - \left(ag + \frac{bg^{n+1}}{n+1}\right)\frac{1}{Dk} + \frac{\omega g}{Dk^2} = c \qquad (12.3.12)$$

We now have a first-order, nonlinear ODE that, upon rearranging, yields

$$dz = \frac{dg}{c + \left(\dfrac{ak - \omega}{Dk^2}\right)g + \dfrac{b}{Dk(n+1)}g^{n+1}}$$

A quadrature is readily derived from this equation by integrating both sides:

$$z = \int \frac{dg}{c + \left(\dfrac{ak - \omega}{Dk^2}\right)g + \dfrac{b}{Dk(n+1)}g^{n+1}} \qquad (12.3.13)$$

The final solution of our original CD equation is obtained by solving the integral and expressing $g(z)$ as a function $u(kx - \omega t)$.

EXERCISE 12.3: Find $u(kx - \omega t)$ using Eq. (12.3.13) and assuming $n = 1$. The nonlinear equation in this case is known as *Burger's equation*.

Integral Transforms

In our discussion of ODE solution techniques, we showed how the Laplace integral could be used to solve an ODE. The Laplace integral is one type of integral transform. Another integral transform is the Fourier integral. Both the Laplace and Fourier integrals can be used to solve some types of PDEs. We demonstrate this technique here using the Fourier integral.

Suppose we want to solve a PDE of the form

$$i\frac{\partial \psi(x, t, s)}{\partial s} = -\left[\frac{\partial^2}{\partial t^2} - \frac{\partial^2}{\partial x^2}\right]\psi(x, t, s) \qquad (12.3.14)$$

for the unknown function $\psi(x, t, s)$ in terms of the independent variables x, t, s. The above equation arises in parameterized relativistic quantum mechanics [Fanchi, 1993] and represents the behavior of a noninteracting, relativistic particle.

The variables $\{x, t, s\}$ correspond to one space coordinate, Einstein's time coordinate, and Newton's time parameter, respectively. Equation (12.3.14) can be solved by trying a solution of the form

$$\psi(x, t, s) = \iiint \eta(k, \omega, \kappa)e^{i\kappa s}e^{i\omega t}e^{-ikx}dk\, d\omega\, d\kappa$$

where the expansion coefficients $\eta(k, \omega, \kappa)$ are determined by as yet unspecified normalization and boundary conditions. The signs of the arguments of the exponential terms were chosen to satisfy relativistic symmetry requirements (Lorentz covariance) so that the solution would have the form of a traveling wave, namely,

$$\psi(x, t, s) = \iiint \eta(k, \omega, \kappa)e^{i\kappa s + i(\omega t - kx)}dk\, d\omega\, d\kappa \qquad (12.3.15)$$

Substituting Eq. (12.3.15) into Eq. (12.3.14) gives

$$\iiint \eta(k, \omega, \kappa)i\frac{\partial}{\partial s}e^{i\kappa s + i(\omega t - kx)}dk\, d\omega\, d\kappa = \iiint \eta(k, \omega, \kappa)\left\{-\left[\frac{\partial^2}{\partial t^2} - \frac{\partial^2}{\partial x^2}\right]\right\}$$
$$\times\, e^{i\kappa s + i(\omega t - kx)}dk\, d\omega\, d\kappa$$

Calculating the derivatives gives

$$\iiint \eta(k, \omega, \kappa)(-\kappa)e^{i\kappa s + i(\omega t - kx)}dk\, d\omega\, d\kappa = \iiint \eta(k, \omega, \kappa)\{-[-\omega^2 - (-k^2)]\}$$
$$\times\, e^{i\kappa s + i(\omega t - kx)}dk\, d\omega\, d\kappa$$

Combining like terms in the preceding equation lets us write

$$\iiint \eta(k, \omega, \kappa)\{[-\kappa - \omega^2 + k^2]\}e^{i\kappa s + i(\omega t - kx)}dk\, d\omega\, d\kappa = 0$$

which is true for all values of k, ω, κ if $\kappa = \omega^2 - k^2$. Thus the Fourier integral representation of the solution ψ leads to a solution of Eq. (12.3.14) with the associated dispersion relation $\kappa = \omega^2 - k^2$. If we now specify normalization and boundary conditions, we can, at least in principle, find explicit expressions for η, ω, k.

EXERCISE 12.4: Find the dispersion relation associated with the Klein–Gordon equation

$$m^2\phi(x, t) = -\left[\frac{\partial^2}{\partial t^2} - \frac{\partial^2}{\partial x^2}\right]\phi(x, t)$$

Hint: Use a solution of the form $\phi(x, t) = \iint \eta(k, \omega)e^{i(\omega t - kx)}dk\, d\omega$.

12.4 NUMERICAL SOLUTION METHODS

Most PDEs, like most ODEs of practical interest, do not lend themselves readily to solution in closed form. It is often necessary to use numerical methods to make progress in constructing a solution of a PDE. Two popular methods of solving PDEs are based on the finite difference method and the finite element method. These methods are discussed here.

Finite Difference Method

We previously showed that numerical solutions of equations with derivatives can be achieved by using a Taylor series to approximate a derivative with a finite difference. The replacement of derivatives with differences can be done in one dimension or in N dimensions. As an illustration, let us consider the one-dimensional CD equation introduced previously:

$$D\frac{\partial^2 u}{\partial x^2} - v\frac{\partial u}{\partial x} = \frac{\partial u}{\partial t} \tag{12.4.1}$$

We assume here that D and v are real, scalar constants. The diffusion term is $D \cdot \partial^2 u/\partial x^2$ and the convection term is $v \cdot \partial u/\partial x$. When the diffusion term is much larger than the convection term, the CD equation behaves like the heat conduction equation, which is a parabolic PDE. If the diffusion term is much smaller than the convection term, the CD equation behaves like a first-order hyperbolic PDE.

The CD equation is especially valuable for studying the implementation of numerical methods for two reasons: (1) the CD equation can be solved analytically and (2) the CD equation may be used to examine two important classes of PDEs (parabolic and hyperbolic). The analytic solution of the CD equation sets the standard for comparing the validity of a numerical method.

To solve the CD equation, we must specify two boundary conditions and an initial condition. We impose the boundary conditions $u(0, t) = 1$, $u(\infty, t) = 0$ for all times t greater than 0, and the initial condition $u(x, 0) = 0$ for all values of x greater than 0. The corresponding solution is [Peaceman, 1977]

$$u(x, t) = \frac{1}{2}\left\{\text{erfc}\left[\frac{x - vt}{2\sqrt{Dt}}\right] + e^{(vx/D)}\text{erfc}\left[\frac{x + vt}{2\sqrt{Dt}}\right]\right\}$$

where the complementary error function $\text{erfc}(y)$ is

$$\text{erfc}(y) = 1 - \frac{2}{\sqrt{\pi}}\int_0^y e^{-z^2} dz$$

Abramowitz and Stegun [1972] have presented an accurate numerical algorithm for calculating $\text{erfc}(y)$.

We now wish to compare the analytic solution of the CD equation with a finite difference representation of the CD equation. If we replace the time and space derivatives with forward and centered differences, respectively, we obtain a discretized form of the CD equation:

$$D \frac{\Delta t}{2\Delta x^2} \left[u_{i+1}^{n+1} - 2u_i^{n+1} + u_{i-1}^{n+1} + u_{i-1}^n - 2u_i^n + u_{i-1}^n \right]$$

$$- v \frac{\Delta t}{\Delta x} \left[\frac{1}{2} \left(u_{i+1}^{n+1} - u_{i-1}^{n+1} \right) \right] = u_i^{n+1} - u_i^n$$

where $\{i, n\}$ denote the discrete levels in the space and time domains, and $\{\Delta x, \Delta t\}$ are step sizes in the respective space and time domains. The finite difference representation of the CD equation leads to a system of linear equations of the form

$$- \left(\frac{v\Delta t}{2\Delta x} + \frac{D\Delta t}{2\Delta x^2} \right) u_{i-1}^{n+1} + \left(1 + \frac{D\Delta t}{\Delta x^2} \right) u_i^{n+1}$$

$$+ \left(\frac{v\Delta t}{2\Delta x} - \frac{D\Delta t}{2\Delta x^2} \right) u_{i+1}^{n+1} = u_i^n + \frac{D\Delta t}{2\Delta x^2} \left(u_{i-1}^n - 2u_i^n + u_{i+1}^n \right)$$

The finite difference equations may be written in matrix form as $Au = B$, where u is the column vector of unknown values $\{u_i^{n+1}\}$, A is a square matrix of coefficients from the left-hand side of the equation, and B is a column vector of terms containing known values $\{u_i^n\}$ from the right-hand side of the equation.

Various methods exist for solving systems of linear equations, but generally these methods fall into one of two groups: (1) direct methods and (2) iterative methods. Direct methods are techniques for solving the system of linear equations in a fixed number of steps. They are capable of solving a wide range of equations, but tend to require large amounts of computer storage. Iterative methods are less robust, but require less storage and are valuable for solving large, sparse systems of equations.

Now that we have both an analytical and a numerical solution of the CD equation, we can make a direct comparison of the methods to demonstrate the validity of the numerical methods. The hyperbolic PDE comparison is shown in Figure 12.1, and the parabolic PDE comparison is shown in Figure 12.2.

These figures show that the finite difference technique does a reasonably good job of reproducing the actual analytical solution. In this case, the parabolic solution is more accurately matched using finite difference methods than is the hyperbolic solution. A more detailed discussion of these methods is beyond the scope of our review. The reader should consult the references for further information, for example, Carnahan et al. [1969], Lapidus and Pinder [1982], Press et al. [1992], and DeVries [1994]. An example of an industrial finite difference simulator is available in Fanchi [1997].

Finite Element Method

Our intent here is to introduce the finite element method by using it to solve a PDE of the form $Lu(\cdot) = f$ for the unknown function $u(\cdot) = u(x, y, z, t)$ given a differential

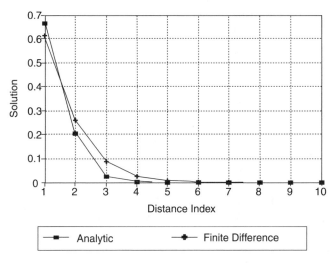

Figure 12.1 Comparison of hyperbolic PDE solutions.

operator L and a function f that is independent of $u(\cdot)$. The finite element method assumes the unknown function $u(\cdot)$ can be approximated by a finite series

$$u(\cdot) \approx \hat{u}(\cdot) = \sum_{j=1}^{N} U_j \phi_j(\cdot)$$

where $\{\phi_j(\cdot)\}$ is a set of basis functions, $\{U_j\}$ are expansion coefficients, and N is the number of terms in the series. The original PDE

$$Lu(\cdot) - f = 0$$

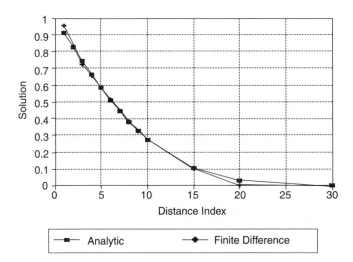

Figure 12.2 Comparison of parabolic PDE solutions.

is replaced by the finite element approximation

$$L\hat{u}(\cdot) - f = R(\cdot)$$

where R is the residual arising from replacing $u(\cdot)$ with $\hat{u}(\cdot)$. In terms of expansion coefficients, the residual is

$$R(\cdot) = L\hat{u}(\cdot) - f = L\left[\sum_{j=1}^{N} U_j \phi_j(\cdot)\right] - f = \sum_{j=1}^{N} U_j \left[L\phi_j(\cdot)\right] - f$$

Our objective is to find the set of expansion coefficients $\{U_j\}$ that minimizes the residual R in the region of interest Ω consisting of the union of the nonoverlapping subdomains $\{V_j t_j\}$ for a given set of basis functions $\{\phi_j(\cdot)\}$. The subdomain $V_j t_j$ is the product of a spatial volume and time in physical applications. To achieve our objective, we invoke the method of weighted residuals.

Method of Weighted Residuals   In this method, N equations are formed by integrating the residual R times a weighting function w_i over the domain of interest Ω; thus

$$\int_t \int_V R(\cdot) w_i(\cdot) \, dV \, dt = 0; i = 1, \ldots, N$$

where $\{w_i\}$ are N specified weighting functions. Inserting the expression for R in terms of expansion coefficients gives the set of integrals

$$\int_t \int_V \left\{ \sum_{j=1}^{N} U_j \left[L\phi_j(\cdot)\right] - f \right\} w_i(\cdot) \, dV \, dt = 0; i = 1, \ldots, N$$

The effectiveness of the method depends to a large extent on selecting an effective set of weighting functions. Some examples are tabulated as follows:

FINITE ELEMENT WEIGHTING FUNCTIONS

Galerkin method	$w_i(\cdot) \equiv \phi_i(\cdot)$
Subdomain method	$w_i(\cdot) \equiv \begin{cases} 1, (x, y, z, t) \in V_i t_i \\ 0, (x, y, z, t) \notin V_i t_i \end{cases}$
Collocation method	$w_i(\cdot) = \delta(x - x_1)$

Example: The best way to breathe life into these ideas is to illustrate their use. Suppose we wish to solve the simple differential equation

$$-K\frac{dp}{dx} = Q$$

for the unknown function $p(x)$ given constants K, Q. The differential operator $-K\,d/dx$ corresponds to L and Q corresponds to the function f. According to the finite element method, we approximate $p(x)$ with the same finite series

$$p(x) \approx \hat{p}(x) = \sum_{j=1}^{N} u_j \phi_j(x)$$

where $\{\phi_j(x)\}$ are basis functions that we specify, and $\{u_j\}$ are the expansion coefficients that we need to compute. The residual is given by $R(\cdot) = L\hat{u}(\cdot) - f$, or

$$R(x) = -K\frac{d\hat{p}}{dx} - Q = -K\sum_{j=1}^{N}\left[u_j\frac{d\phi_j}{dx}\right] - Q$$

Our task is to find the expansion coefficients (u_j) that minimize the residual in the region of interest $\Omega = \{x : 0 \leq x \leq 1\}$ using the method of weighted residuals.

Illustrative Basis Functions   Some typical basis functions are the straight-line segments

$$\phi_1(x) = \begin{cases} -2x + 1, & 0 \leq x \leq 0.5 \\ 0, & 0.5 \leq x \leq 1.0 \end{cases}$$

$$\phi_2(x) = \begin{cases} 2x, & 0 \leq x \leq 0.5 \\ -2x + 2, & 0.5 \leq x \leq 1.0 \end{cases}$$

$$\phi_3(x) = \begin{cases} 0, & 0 \leq x \leq 0.5 \\ 2x - 1, & 0.5 \leq x \leq 1.0 \end{cases}$$

in the domain of interest for a finite series with $N = 3$ basis functions. These functions are sketched in Figure 12.3. The function $\phi_1(x)$ provides a nonzero contribution for the interval between nodes 1 and 2. Function $\phi_3(x)$ serves a similar role in the interval between nodes 2 and 3.

Applying the Method of Weighted Residuals   To apply the method of weighted residuals, we must solve the set of N equations

$$\int_{\Omega} R(x)w_i(x)\,dx = 0; \quad i = 1, \ldots, N$$

for the expansion coefficients $\{u_j\}$. For our illustration, we have

$$\int_0^1 \left\{ K\sum_{j=1}^{N}\left[u_j\frac{d\phi_j}{dx}\right] + Q \right\} w_i(x)\,dx = 0$$

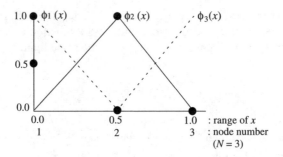

Figure 12.3 Finite element functions.

where $\{w_i\}$ is the set of specified weighting functions. Several choices of weighting functions are possible. Two of the most common are the Galerkin method and the collocation method. The weighting functions for each method are the following:

$$\text{Gelerkin method:} \quad w_i(x) = \phi_i(x); \text{ basis function}$$
$$\text{Collocation method:} \quad w_i(x) = \delta(x - x_i); \text{ delta function}$$

Actual solution of the problem from this point would require performing the integrals and then solving for the expansion coefficients. This is a tedious process best suited for computers. For further discussion of the finite element method, consult such references as Lapidus and Pinder [1982] or Press et al. [1992].

CHAPTER 13

INTEGRAL EQUATIONS

Equations in which the unknown function is a factor in the integrand and may be a term outside the integral are called integral equations. They arise in such disciplines as nuclear reactor engineering; the study of fluid flow in fractured, porous media; and the description of particle scattering. We present below a classification scheme for integral equations and three solution techniques. Arfken and Weber [2001], Corduneanu [1991], Collins [1999], and Chow [2000] present additional discussion of integral equations.

13.1 CLASSIFICATION

Integral equations are classified according to their integration limits and whether or not the unknown function appears outside the integrand. A summary of classification terms is as follows:

INTEGRAL EQUATION TERMINOLOGY

Fredholm equation	Both integration limits are fixed
Volterra equation	One limit of integration is variable
Integral equation of the first kind	Unknown function appears only as part of the integrand
Integral equation of the second kind	Unknown function appears inside and outside the integral

Math Refresher for Scientists and Engineers, Third Edition By John R. Fanchi
Copyright © 2006 John Wiley & Sons, Inc.

Example: Suppose the kernel $K(x, t)$ and the function $f(x)$ are known, and $\phi(t)$ is the unknown function. Examples of integral equations are as follows:

EXAMPLES OF INTEGRAL EQUATIONS

Fredholm equation of the first kind	$f(x) = \int_a^b K(x, t)\, \phi(t)dt$
Fredholm equation of the second kind	$\phi(x) = f(x) + \lambda \int_a^b K(x, t)\, \phi(t)dt$
Volterra equation of the first kind	$f(x) = \int_a^x K(x, t)\, \phi(t)dt$
Volterra equation of the second kind	$\phi(x) = f(x) + \lambda \int_a^x K(x, t)\, \phi(t)dt$

Note: If $f(x) = 0$, the integral equation is homogeneous.

13.2   INTEGRAL EQUATION REPRESENTATION OF A SECOND-ORDER ODE

An important class of integral equations is obtained by transforming the second-order differential equation

$$y'' + A(x)y' + B(x)y = g(x) \tag{13.2.1}$$

for the unknown function $y(x)$ into an integral equation. The unknown function satisfies the initial conditions

$$y(a) = y_0, y'(a) = y_0' \tag{13.2.2}$$

at the point $x = a$. We first integrate Eq. (13.2.1) to find

$$y' = -\int_a^x A(x)y'dx - \int_a^x B(x)ydx + \int_a^x g(x)dx + y_0' \tag{13.2.3}$$

where y_0' is an integration constant and we have used the relation

$$\int_a^x \frac{d^2y}{dx^2}dx = \int_a^x \frac{d}{dx}\left(\frac{dy}{dx}\right)dx = \int_a^x d\left(\frac{dy}{dx}\right) = y'(x) - y'(a) = y'(x) - y_0'$$

Now integrate the first integral of Eq. (13.2.3) by parts to get

$$\int_a^x u\,dv = uv\big|_a^x - \int_a^x v\,du$$

where

$$u = A(x), dv = y'dx = dy$$

The result is

$$-\int_a^x A(x)y'dx = -A(x)\,y(x)\Big|_a^x + \int_a^x y\,dA = -Ay + A(a)\,y(a) + \int_a^x y\frac{dA}{dx}dx \quad (13.2.4)$$

Writing $A' = dA/dx$ and using Eq. (13.2.4) in Eq. (13.2.3) yields

$$y' = -Ay - \int_a^x (B - A')y\,dx + \int_a^x g(x)dx + A(a)\,y(a) + y_0' \quad (13.2.5)$$

where $y(a) = y_0$.

Integrating both sides of Eq. (13.2.5) lets us write

$$\int_a^x y'dx = -\int_a^x Ay\,dx - \int_a^x \left\{\int_a^x [B(t) - A'(t)]\,y(t)dt\right\}dx$$

$$+ \int_a^x \left\{\int_a^x g(t)dt\right\}dx + \int_a^x A(a)\,y(a)dx + \int_a^x y_0'\,dx \quad (13.2.6)$$

where t is a dummy index. The last integral on the right-hand side of Eq. (13.2.6) is

$$\int_a^x y_0'dx = y_0'\int_a^x dx = y_0'(x - a)$$

The integral on the left-hand side of Eq. (13.2.6) is evaluated using the initial condition $y(a) = y_0$ to yield

$$\int_a^x y'dx = y\Big|_a^x = y(x) - y(a) = y(x) - y_0$$

Substituting the above integrals into Eq. (13.2.6) gives

$$y(x) = -\int_a^x Aydx - \int_a^x \left\{\int_a^x [B(t) - A'(t)]\,y(t)dt\right\}dx$$

$$+ \int_a^x \left\{\int_a^x g(t)dt\right\}dx + [A(a)y_0 + y_0'](x - a) + y_0 \quad (13.2.7)$$

The two integrals with curly brackets $\{\cdots\}$ are simplified using the relation [Arfken and Weber, 2001, p. 986]

$$\int_a^x \left\{\int_a^x f(t)dt\right\}dx = \int_a^x \left\{\int_a^x f(t)dx\right\}dt = \int_a^x (x - t)f(t)dt \quad (13.2.8)$$

Equation (13.2.8) can be verified by differentiating both sides with respect to x. The equality of the resulting derivatives implies that the original expressions in

Eq. (13.2.8) can only differ by a constant. Letting the limit $x \to a$ so that $\int_a^x \to \int_a^a$ implies the equality $0 = 0$ holds and the constant must equal 0. Substituting Eq. (13.2.8) into Eq. (13.2.7) gives

$$y(x) = -\int_a^x Ay\,dx - \int_a^x (x-t)[B(t) - A'(t)]\,y(t)dt$$

$$+ \int_a^x (x-t)\,g(t)dt + [A(a)y_0 + y_0'](x-a) + y_0$$

Collecting terms and simplifying lets us write

$$y(x) = -\int_a^x \big\{A(t)\,y(t) + (x-t)[B(t) - A'(t)]\,y(t)\big\}dt$$

$$+ \int_a^x (x-t)\,g(t)dt + [A(a)y_0 + y_0'](x-a) + y_0$$

If we define the kernel $K(x, t)$ as

$$K(x,t) = (t-x)[B(t) - A'(t)] - A(t) \qquad (13.2.9)$$

and the function

$$f(x) = \int_a^x (x-t)g(t)dt + [A(a)y_0 + y_0'](x-a) + y_0 \qquad (13.2.10)$$

then we can write the formal solution to the second-order ODE (13.2.1) in the form

$$y(x) = f(x) + \int_a^x K(x,t)y(t)dt \qquad (13.2.11)$$

Equation (13.2.11) is a Volterra equation of the second kind.

Application: Linear Oscillator. The linear oscillator equation $y'' + \omega^2 y = 0$ has many real-world applications and can be solved as an ordinary differential equation. It can also be represented as an integral equation by comparing $y'' + \omega^2 y = 0$ to Eq. (13.2.1) and recognizing that

$$A(x) = 0, B(x) = \omega^2, g(x) = 0 \qquad (13.2.12)$$

Substituting Eq. (13.2.12) into Eqs. (13.2.9) and (13.2.10) gives the kernel

$$K(x,t) = (t-x)[B(t) - A'(t)] - A(t) = (t-x)\omega^2 \qquad (13.2.13)$$

and the function

$$f(x) = [y_0'](x-a) + y_0 \qquad (13.2.14)$$

The formal solution of the linear oscillator equation $y'' + \omega^2 y = 0$ is

$$y(x) = f(x) + \int_a^x K(x,t)y(t)dt = y_0'(x-a) + y_0 + \omega^2 \int_a^x (t-x)y(t)dt \qquad (13.2.15)$$

where the initial conditions at the point $x = a$ are $y(a) = y_0, y'(a) = y_0'$.

EXERCISE 13.1: Write the linear oscillator equation $y'' + \omega^2 y = 0$ with initial conditions $y(0) = 0, y'(0) = 1$ in the form of an integral equation.

EXERCISE 13.2: Solve the linear oscillator equation $y'' + \omega^2 y = 0$ with initial conditions $y(0) = 0, y(b) = 0$. *Hint*: Modify the above procedure to account for the use of boundary conditions $y(0) = 0, y(b) = 0$ [see Arfken and Weber, 2001, pp. 987–989].

13.3 SOLVING INTEGRAL EQUATIONS: NEUMANN SERIES METHOD

We would like to find a solution of an integral equation such as the inhomogeneous Fredholm equation

$$\phi(x) = f(x) + \lambda \int_a^b K(x,t)\,\phi(t)dt \qquad (13.3.1)$$

where $f(x) \neq 0$. Assume the integral in Eq. (13.3.1) can be treated as a perturbation of $f(x)$ so that we can try the solution

$$\phi(x) \approx \phi_0(x) = f(x) \qquad (13.3.2)$$

as a zero-order approximation. The first-order approximation of the solution $\phi(x)$ is obtained by substituting Eq. (13.3.2) into Eq. (13.3.1); thus

$$\phi_1(x) = f(x) + \lambda \int_a^b K(x,t)f(t)dt \qquad (13.3.3)$$

We now proceed iteratively and use $\phi(x) \approx \phi_1(x)$ in Eq. (13.3.1) to obtain the second-order approximation

$$\phi_2(x) = f(x) + \lambda \int_a^b K(x,t_1)f(t_1)dt_1 + \lambda^2 \int_a^b \int_a^b K(x,t_1)K(t_1,t_2)f(t_2)dt_1 dt_2 \qquad (13.3.4)$$

Rearranging Eq. (13.3.4) and combining terms gives

$$\phi_2(x) = f(x) + \lambda \int_a^b K(x, t_1) \left[f(t_1) + \lambda \int_a^b K(t_1, t_2) f(t_2) \, dt_1 \right] dt_1 \qquad (13.3.5)$$

For the nth order approximation we have

$$\phi_n(x) = \sum_{i=0}^n \lambda^i u_i(x) \qquad (13.3.6)$$

where

$$u_0(x) = f(x)$$

$$u_1(x) = \int_a^b K(x, t_1) f(t_1) \, dt_1$$

$$u_2(x) = \int_a^b \int_a^b K(x, t_1) K(t_1, t_2) f(t_2) \, dt_2 \, dt_1 \qquad (13.3.7)$$

$$\vdots$$

$$u_n(x) = \int_a^b \cdots \int_a^b K(x, t_1) K(t_1, t_2) \cdots K(t_{n-1}, t_n) f(t_n) dt_n \cdots dt_1$$

The solution of Eq. (13.3.1) is

$$\phi(x) = \lim_{n \to \infty} \phi_n(x) = \lim_{n \to \infty} \sum_{i=0}^n \lambda^i u_i(x) \qquad (13.3.8)$$

if the series in Eq. (13.3.8) converges. Eq. (13.3.8) is the Neumann series solution of the inhomogeneous Fredholm equation given by Eq. (13.3.1).

An analogous procedure can be used to find a solution of the inhomogeneous Volterra equation

$$\phi(x) = f(x) + \lambda \int_a^x K(x, t) \phi(t) \, dt \qquad (13.3.9)$$

The zero-order approximation is

$$\phi(x) \approx \phi_0(x) = f(x) \qquad (13.3.10)$$

Substituting Eq. (13.3.10) into Eq. (13.3.9) gives the first-order approximation

$$\phi_1(x) = f(x) + \lambda \int_a^x K(x, t) f(t) \, dt \qquad (13.3.11)$$

For the second-order approximation, let $\phi(x) \approx \phi_1(x)$ to find

$$\phi_2(x) = f(x) + \lambda \int_a^x K(x, t_1)\,\phi(t_1)dt_1$$

$$= f(x) + \lambda \int_a^x K(x, t_1)\left[f(t_1) + \lambda \int_a^{t_1} K(t_1, t_2)f(t_2)dt_2 \right]dt_1$$

(13.3.12)

where $\phi(t_1) \approx \phi_1(t_1)$. Expanding the bracketed term gives

$$\phi_2(x) = f(x) + \lambda \int_a^x K(x, t_1)f(t_1)dt_1 + \lambda^2 \int_a^x \int_a^{t_1} K(x, t_1)K(t_1, t_2)f(t_2)\,dt_2\,dt_1 \quad (13.3.13)$$

Higher-order approximations may be obtained iteratively by repeating the procedure outlined above.

13.4 SOLVING INTEGRAL EQUATIONS WITH SEPARABLE KERNELS

Integral equations with separable kernels may be solved using the following procedure. A kernel $K(x, t)$ is separable if it can be written in the form

$$K(x, t) = \sum_{j=1}^n M_j(x)\,N_j(t) \tag{13.4.1}$$

where n is finite, and $M_j(x), N_j(t)$ are polynomials or elementary transcendental functions. Substituting Eq. (13.4.1) into the Fredholm equation of the second kind given by Eq. (13.3.1) gives

$$\phi(x) = f(x) + \lambda \int_a^b K(x,t)\phi(t)dt = f(x) + \lambda \sum_{j=1}^n \int_a^b M_j(x)\,N_j(t)\,\phi(t)dt \tag{13.4.2}$$

Equation (13.4.2) can be rewritten as

$$\phi(x) = f(x) + \lambda \sum_{j=1}^n M_j(x)c_j \tag{13.4.3}$$

where

$$c_j = \int_a^b N_j(t)\,\phi(t)dt \tag{13.4.4}$$

are numbers. To determine c_j, multiply Eq. (13.4.3) by $N_i(x)$ and integrate:

$$\int_a^b N_i(x)\,\phi(x)dx = \int_a^b N_i(x)f(x)dx + \lambda \sum_{j=1}^n \int_a^b N_i(x)\,M_j(x)c_j\,dx \tag{13.4.5}$$

Equation (13.4.5) has the form

$$\int_a^b N_i(x)\phi(x)dx = c_i = b_i + \lambda \sum_{j=1}^n a_{ij}c_j \tag{13.4.6}$$

where we have used Eq. (13.4.4) and the definitions

$$b_i = \int_a^b N_i(x)f(x)dx \tag{13.4.7}$$

and

$$a_{ij}c_j = \int_a^b N_i(x)M_j(x)c_jdx \tag{13.4.8}$$

The constants $\{c_j\}$ can be factored out of Eq. (13.4.8) to yield

$$a_{ij} = \int_a^b N_i(x)M_j(x)\,dx \tag{13.4.9}$$

Equation (13.4.6) can be written in matrix form as

$$c - \lambda Ac = b = (I - \lambda A)c \tag{13.4.10}$$

where I is the identity matrix. Equation (13.4.10) has the formal solution

$$c = (I - \lambda A)^{-1}b \tag{13.4.11}$$

If the index n in Eq. (13.4.1) is finite, the set of equations represented by Eqs. (13.4.10) and (13.4.11) are finite. Solving the determinant $|I - \lambda A| = 0$ gives a set of n eigenvalues $\{\lambda_j\}$. Substituting the resulting eigenvalues into Eq. (13.4.10) lets us calculate $\{c_j\}$. Given the set $\{c_j\}$, we solve Eq. (13.4.3) by using a trial and error procedure to determine the functions $M_j(x)$, $N_j(t)$.

EXERCISE 13.3: Solve the homogeneous Fredholm equation $\phi(x) = \lambda \int_{-1}^1 (t + x)\,\phi(t)dt$ [see Arfken and Weber, 2001, p. 1002].

13.5 SOLVING INTEGRAL EQUATIONS WITH LAPLACE TRANSFORMS

The Laplace transform of an integral presented in Section 9.6 can be used to solve some types of integral equations. Consider the Volterra equation of the second kind,

$$\phi(t) = f(t) + \lambda \int_0^t \phi(\tau)d\tau \tag{13.5.1}$$

where $f(t)$ is an arbitrary function of t, the lower limit of the integral is set to zero, and the kernel is $K(t, \tau) = 1$. The Laplace transform of Eq. (13.5.1) is

$$\mathcal{L}\{\phi(t)\} = \mathcal{L}\left\{f(t) + \lambda \int_0^t \phi(\tau)d\tau\right\}$$

$$= \mathcal{L}\{f(t)\} + \mathcal{L}\left\{\lambda \int_0^t \phi(\tau)d\tau\right\} \quad (13.5.2)$$

$$= \mathcal{L}\{f(t)\} + \lambda\mathcal{L}\left\{\int_0^t \phi(\tau)d\tau\right\}$$

The Laplace transform of the integral is given by Eq. (9.6.12) so that Eq. (13.5.2) becomes

$$\mathcal{L}\{\phi(t)\} = \mathcal{L}\{f(t)\} + \lambda\frac{\mathcal{L}\{\phi(t)\}}{s} \quad (13.5.3)$$

where s is a real, positive parameter. Solving for $\mathcal{L}\{\phi(t)\}$ gives

$$\mathcal{L}\{\phi(t)\} = \frac{\mathcal{L}\{f(t)\}}{1 - \dfrac{\lambda}{s}} = \frac{s\mathcal{L}\{f(t)\}}{s - \lambda} \quad (13.5.4)$$

Taking the inverse Laplace transform of Eq. (13.5.4) gives the formal solution

$$\phi(t) = \mathcal{L}^{-1}\left\{\frac{s}{s - \lambda}\mathcal{L}\{f(t)\}\right\} \quad (13.5.5)$$

EXERCISE 13.4: Use the Laplace transform to solve the integral equation

$$V_0 \cos \omega t = Ri(t) + \frac{1}{C}\int_0^t i(\tau)\,d\tau$$

for an RC circuit. An RC circuit consists of a resistor with resistance R and a capacitor with capacitance C in series with a time-varying voltage $V(t) = V_0 \cos \omega t$ that has frequency ω.

CHAPTER 14

CALCULUS OF VARIATIONS

The calculus of variations provides a methodology for finding a path of integration that satisfies an extremum condition. In this chapter, we present the calculus of variations for a single dependent variable, for several dependent variables, and the calculus of variations with constraints.

14.1 CALCULUS OF VARIATIONS WITH ONE DEPENDENT VARIABLE

The path of integration of a line integral was defined in Chapter 9 as a smooth curve C that may be written as the vector

$$\mathbf{r}(s) = x(s)\hat{i} + y(s)\hat{j} + z(s)\hat{k} \qquad (14.1.1)$$

in three dimensions where the arc length s parameterizes the coordinates between the end points $\{s = a, \ s = b\}$. The vector $\mathbf{r}(s)$ is illustrated in Figure 9.1. It is continuous and differentiable when the path of integration is a smooth curve.

The calculus of variations can be used to find a path of integration $C(\alpha)$ that makes the line integral $J(\alpha)$ an extremum with respect to a parameter α. Let us choose the path of integration $C(\alpha = 0)$ as the unknown path of integration that

Math Refresher for Scientists and Engineers, Third Edition By John R. Fanchi
Copyright © 2006 John Wiley & Sons, Inc.

minimizes $J(\alpha)$. The condition for finding $C(\alpha = 0)$ given $J(\alpha)$ is

$$\frac{\partial J(\alpha)}{\partial \alpha}\bigg|_{\alpha=0} = 0 \qquad (14.1.2)$$

where the line integral is considered to be a function of the parameter α.

We illustrate the calculus of variations by considering the line integral J_1 of a function $f_1(x, \dot{x}, t)$ that depends on the one-dimensional path x and its derivative $\dot{x} = dx/dt$ with respect to a variable t. The function f_1, the path x, the derivative of the path $\dot{x}$, and the line integral J_1 may all depend on a parameter α that parameterizes the path of integration. The parameterized line integral may be written as

$$J_1(\alpha) = \int_{t_1}^{t_2} f_1(x(t, \alpha), \ \dot{x}(t, \alpha), t) dt \qquad (14.1.3)$$

where the path $x = x(t, \alpha)$ is defined between the end points $\{t_1, t_2\}$.

The path of integration that corresponds to an extremum of J_1 is the path $x = x(t, \alpha)$ that satisfies Eq. (14.1.2). For example, we may write the path of integration as

$$x(t, \ \alpha) = x(t, \ 0) + \varepsilon(\alpha)\eta(t) \qquad (14.1.4)$$

The path $x(t, 0)$ is the path that satisfies Eq. (14.1.2) and the function $\varepsilon(\alpha)\eta(t)$ is the deformation of the path relative to the path $x(t, 0)$. The paths are illustrated in Figure 14.1. The deformation $\varepsilon(\alpha)\eta(t)$ must equal zero at the end points $\{t_1, t_2\}$ to ensure that all parameterized paths pass through the end points.

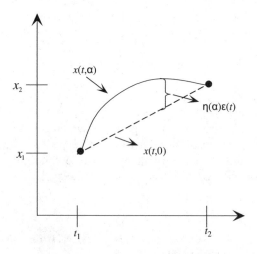

Figure 14.1 Paths of integration.

Applying the extremum condition in Eq. (14.1.2) to Eq. (14.1.3) gives

$$
\begin{aligned}
\frac{\partial J_1(\alpha)}{\partial \alpha} &= \frac{\partial}{\partial \alpha} \int_{t_1}^{t_2} f_1(x(t, \alpha), \dot{x}(t, \alpha), t)\, dt \\
&= \int_{t_1}^{t_2} \left[\frac{\partial f_1}{\partial x} \frac{\partial x}{\partial \alpha} + \frac{\partial f_1}{\partial \dot{x}} \frac{\partial \dot{x}}{\partial \alpha} \right] dt = 0
\end{aligned}
\tag{14.1.5}
$$

The second integral on the right-hand side of Eq. (14.1.5) can be integrated by parts to yield

$$
\begin{aligned}
\int_{t_1}^{t_2} \frac{\partial f_1}{\partial \dot{x}} \frac{\partial \dot{x}}{\partial \alpha}\, dt &= \int_{t_1}^{t_2} \frac{\partial f_1}{\partial \dot{x}} \frac{\partial^2 x}{\partial t\, \partial \alpha}\, dt \\
&= \frac{\partial f_1}{\partial \dot{x}} \frac{\partial x}{\partial \alpha} \Big|_{t_1}^{t_2} - \int_{t_1}^{t_2} \frac{d}{dt}\left(\frac{\partial f_1}{\partial \dot{x}} \right) \frac{\partial x}{\partial \alpha}\, dt
\end{aligned}
\tag{14.1.6}
$$

The partial derivative $\partial x / \partial \alpha$ is arbitrary except at the end points $\{t_1, t_2\}$, where $\partial x / \partial \alpha$ must equal zero to guarantee that all of the parameterized paths of integration pass through the points (t_1, x_1), (t_2, x_2) for any value of the parameter α. Consequently, the first term on the right-hand side of Eq. (14.1.6) vanishes (see Exercise 14.1) and Eq. (14.1.6) simplifies to

$$
\int_{t_1}^{t_2} \frac{\partial f_1}{\partial \dot{x}} \frac{\partial \dot{x}}{\partial \alpha}\, dt = -\int_{t_1}^{t_2} \frac{d}{dt}\left(\frac{\partial f_1}{\partial \dot{x}} \right) \frac{\partial x}{\partial \alpha}\, dt
\tag{14.1.7}
$$

Substituting Eq. (14.1.7) into Eq. (14.1.5) gives

$$
\begin{aligned}
\frac{\partial J_1(\alpha)}{\partial \alpha} &= \int_{t_1}^{t_2} \left[\frac{\partial f_1}{\partial x} \frac{\partial x}{\partial \alpha} - \frac{d}{dt}\left(\frac{\partial f_1}{\partial \dot{x}} \right) \frac{\partial x}{\partial \alpha} \right] dt \\
&= \int_{t_1}^{t_2} \left[\frac{\partial f_1}{\partial x} - \frac{d}{dt}\left(\frac{\partial f_1}{\partial \dot{x}} \right) \right] \frac{\partial x}{\partial \alpha}\, dt \\
&= 0
\end{aligned}
\tag{14.1.8}
$$

The extremum condition $(\partial J_1(\alpha)/\partial \alpha)|_{\alpha=0} = 0$ is satisfied when the integrand in Eq. (14.1.8) vanishes at $\alpha = 0$. The integrand vanishes for an arbitrary deformation $\partial x / \partial \alpha$ when the term in brackets vanishes; therefore the extremum condition $(\partial J_1(\alpha)/\partial \alpha)|_{\alpha=0} = 0$ is satisfied when

$$
\frac{\partial f_1}{\partial x} - \frac{d}{dt}\left(\frac{\partial f_1}{\partial \dot{x}} \right) = 0
\tag{14.1.9}
$$

Equation (14.1.9) is known as the Euler–Lagrange equation.

EXERCISE 14.1: Show that Eq. (14.1.4) implies that $(\partial x/\partial \alpha)|_{t_1}^{t_2} = 0$.

EXERCISE 14.2: What is the shortest distance between two points in a plane? *Hint*: The element of arc length in a plane is $ds = \sqrt{dx^2 + dy^2} = \sqrt{1 + (dy/dx)^2} \, dx$.

Application: Classical Dynamics. Joseph Louis Lagrange used the calculus of variations to formulate the dynamical laws of motion in terms of a function of energy. We illustrate Lagrange's formulation in one space dimension by introducing a function L that depends on the coordinate q and its velocity $\dot{q} = dq/dt$, where t is time. The function L is called the Lagrangian and is the difference between kinetic energy T and potential energy V. Kinetic energy T may depend on position and velocity. Potential energy V may depend on position only for a conservative force, that is, a force that can be derived from the gradient of the potential energy. The resulting Lagrangian has the form

$$L = L(q, \dot{q}) = T(q, \dot{q}) - V(q) \qquad (14.1.10)$$

Lagrange's equations are derived from William R. Hamilton's principle of least action. Action S is defined as the integral

$$S = \int_{t_1}^{t_2} L(q, \dot{q}, t) \, dt \qquad (14.1.11)$$

between times $\{t_1, t_2\}$. The dependence of the Lagrangian on time is shown explicitly in Eq. (14.1.11). Hamilton's principle of least action is a variational principle. It says that a classical particle will follow a path that makes the action integral an extremum. If we introduce the parameter α, Eq. (14.1.11) becomes

$$S(\alpha) = \int_{t_1}^{t_2} L(q(t, \alpha), \ \dot{q}(t, \alpha), t) \, dt \qquad (14.1.12)$$

Equation (14.1.12) has the same form as the path integral in Eq. (14.1.3). The parameterized paths of integration corresponding to Eq. (14.1.4) may be written as

$$q(t, \alpha) = q(t, 0) + \varepsilon(\alpha)\eta(t) \qquad (14.1.13)$$

Following the derivation above, we obtain the Euler–Lagrange equation

$$\frac{\partial L}{\partial q} - \frac{d}{dt}\left(\frac{\partial L}{\partial \dot{q}}\right) = 0 \qquad (14.1.14)$$

Momentum p is calculated from the Lagrangian as

$$p = \frac{\partial L}{\partial \dot{q}} \qquad (14.1.15)$$

and force F is

$$F = \frac{\partial L}{\partial q} \tag{14.1.16}$$

Example: We illustrate the range of applicability of the calculus of variations by considering an LC circuit. An LC circuit consists of a capacitor and an inductor in series. In this example, the capacitor has constant capacitance C and the inductor has constant inductance L. The Lagrangian for the LC circuit is

$$L = L(q, \dot{q}) = T(q, \dot{q}) - V(q) = \frac{1}{2} L\dot{q}^2 - \frac{1}{2}\frac{q^2}{C} \tag{14.1.17}$$

where charge q is a generalized coordinate that depends on the independent variable time t. Electric current $I = \dot{q}$ is the time rate of change of charge. The energy $L\dot{q}^2/2$ associated with the inductor corresponds to the kinetic energy term, and the energy $q^2/(2C)$ stored in the capacitor corresponds to the potential energy term. The voltage drop around the LC circuit can be determined by substituting Eq. (14.1.17) into the Euler–Lagrange equation. The resulting differential equation is

$$\frac{\partial L}{\partial q} - \frac{d}{dt}\left(\frac{\partial L}{\partial \dot{q}}\right) = -\frac{q}{C} - \frac{d}{dt}(L\dot{q}) = -\frac{q}{C} - L\ddot{q} = 0$$

or

$$L\ddot{q} + \frac{q}{C} = 0 \tag{14.1.18}$$

Equation (14.1.18) is the voltage drop equation for an LC circuit.

EXERCISE 14.3: The kinetic energy and potential energy of a harmonic oscillator (HO) in one space dimension are $T_{HO} = \frac{1}{2}m\dot{q}^2$ and $V_{HO} = \frac{1}{2}mq^2$, respectively, where m is the constant mass of the oscillating object and k is the spring constant. Write the Lagrangian for the harmonic oscillator and use it in the Euler–Lagrange equation to determine the force equation for the harmonic oscillator.

14.2 THE BELTRAMI IDENTITY AND THE BRACHISTOCHRONE PROBLEM

A problem that motivated the development of the calculus of variations was the brachistochrone or shortest time problem. The goal of the brachistochrone problem is to determine the shape of a frictionless wire that minimizes the time of descent of a bead sliding along the wire between two points in a vertical plane under the influence of gravity. The bead has constant mass m and speed v. Its time of descent is given by the integral

$$t_{12} = \int_{s_1}^{s_2} \frac{ds}{v} \tag{14.2.1}$$

where s is the arc length along the wire in a two-dimensional $\{x, z\}$ plane with end points $\{s_1(x_1, z_1), s_2(x_2, z_2)\}$. The element of arc length is

$$ds = \sqrt{dx^2 + dz^2} = \left(\sqrt{1 + \left(\frac{dz}{dx}\right)^2} \right) dx = \sqrt{1 + z'^2}\, dx \tag{14.2.2}$$

where we are using the notation $z' = dz/dx$. If we assume the elevation $z = 0$ at the point of release of the bead, we can determine the speed of the bead from conservation of energy $\frac{1}{2}mv^2 = mgz$ to be

$$v = \sqrt{2gz} \tag{14.2.3}$$

We change the variable of integration in Eq. (14.2.1) using Eq. (14.2.2) and write the speed of the bead using Eq. (14.2.3) to find

$$t_{12} = \int_{z_1}^{z_2} \frac{\sqrt{1 + z'^2}}{\sqrt{2gz}}\, dx \tag{14.2.4}$$

Equation (14.2.4) has the form of Eq. (14.1.3) with

$$f_1(z(x, \alpha), \dot{z}'(x, \alpha), x) = \sqrt{\frac{1 + \dot{z}'^2}{\sqrt{2gz}}} \tag{14.2.5}$$

The corresponding Euler–Lagrange equation is

$$\frac{\partial f_1}{\partial z} - \frac{d}{dx}\left(\frac{\partial f_1}{\partial z'}\right) = 0 \tag{14.2.6}$$

We could attempt to solve the brachistochrone problem by substituting Eq. (14.2.5) into Eq. (14.2.6) and solving the resulting differential equation. An alternative approach that is more direct for this problem is to use an alternative form of the Euler–Lagrange equation that is known as the Beltrami identity.

The Beltrami identity is obtained by considering the differential of a function of the form $f(z(x, \alpha), z'(x, \alpha), x)$, thus

$$\frac{df}{dx} = \frac{\partial f}{\partial x} + \frac{\partial f}{\partial z}\frac{dz}{dx} + \frac{\partial f}{\partial z'}\frac{dz'}{dx} \tag{14.2.7}$$

We now solve Eq. (14.2.7) for the term containing $\partial f/\partial z$:

$$\frac{\partial f}{\partial z}\frac{dz}{dx} = \frac{df}{dx} - \frac{\partial f}{\partial x} - \frac{\partial f}{\partial z'}\frac{dz'}{dx} \tag{14.2.8}$$

If we multiply the Euler–Lagrange equation

$$\frac{\partial f}{\partial z} - \frac{d}{dx}\left(\frac{\partial f}{\partial z'}\right) = 0$$

for $f(z(x, \alpha), z'(x, \alpha), x)$ by dz/dx we obtain

$$\frac{\partial f}{\partial z}\frac{dz}{dx} - \frac{dz}{dx}\frac{d}{dx}\left(\frac{\partial f}{\partial z'}\right) = 0 \tag{14.2.9}$$

Substituting Eq. (14.2.8) into Eq. (14.2.9) gives

$$\frac{df}{dx} - \frac{\partial f}{\partial x} - \frac{\partial f}{\partial z'}\frac{dz'}{dx} - \frac{dz}{dx}\frac{d}{dx}\left(\frac{\partial f}{\partial z'}\right) = 0 \tag{14.2.10}$$

Combining terms in Eq. (14.2.10) gives the Beltrami identity

$$-\frac{\partial f}{\partial x} + \frac{d}{dx}\left(f - \frac{dz}{dx}\frac{\partial f}{\partial z'}\right) = 0 \tag{14.2.11}$$

Equation (14.2.11) is especially useful when $\partial f/\partial x = 0$ since

$$\frac{d}{dx}\left(f - \frac{dz}{dx}\frac{\partial f}{\partial z'}\right) = 0 \tag{14.2.12}$$

which can be integrated to yield

$$f - \frac{dz}{dx}\frac{\partial f}{\partial z'} = C \tag{14.2.13}$$

where C is a constant of integration.

The Beltrami identity, especially Eq. (14.2.13), can be used to solve the brachistochrone problem because f_1 in Eq. (14.2.5) satisfies $\partial f_1/\partial x = 0$. Substituting Eq. (14.2.5) into the Beltrami identity and observing that $\partial f_1/\partial x = 0$ gives

$$f_1 - \frac{dz}{dx}\frac{\partial f_1}{\partial z'} = C \tag{14.2.14}$$

Evaluating derivatives in Eq. (14.2.14) and simplifying lets us write

$$\sqrt{\frac{1 + z'^2}{2gz}} - z'^2\sqrt{\frac{1}{2gz(1 + z'^2)}} = C \tag{14.2.15}$$

since

$$\frac{\partial f_1}{\partial z'} = z' \sqrt{\frac{1}{2gz(1 + z'^2)}} \tag{14.2.16}$$

Combining terms in Eq. (14.2.15) yields

$$\sqrt{\frac{1}{z(1 + z'^2)}} = \sqrt{2g}C = \frac{1}{k} \tag{14.2.17}$$

where we have redefined the constant of integration as $k^{-1} = \sqrt{2g}C$. Solving Eq. (14.2.17) for z' gives

$$z' = \frac{dz}{dx} = \sqrt{\frac{k^2 - z}{z}}$$

or

$$\frac{dz}{dx} = \sqrt{\frac{a - z}{z}} \tag{14.2.18}$$

where $a = k^2$. The solution to Eq. (14.2.18) that solves the brachistochrone problem is a cycloid.

EXERCISE 14.4: Solve Eq. (14.2.18). *Hint*: Perform a variable substitution by introducing a parameter θ that lets you substitute $z = a \sin^2 \theta = (a/2)(1 - \cos 2\theta)$ for z.

14.3 CALCULUS OF VARIATIONS WITH SEVERAL DEPENDENT VARIABLES

The calculus of variations introduced in Section 14.1 for one dependent variable is extended to several dependent variables in this section. We illustrate the calculus of variations with several dependent variables by considering the line integral J_n of a function $f_n(x_1, x_2, \ldots, x_n, \dot{x}_1, \dot{x}_2, \ldots, \dot{x}_n, t)$ that depends on n dependent variables $\{x_i = x_i(t), i = 1, \ldots, n\}$ and their derivatives $\{\dot{x}_i = dx_i/dt, i = 1, \ldots, n\}$ with respect to an independent variable t. The function f_n, the path $\{x_i = x_i(t), i = 1, \ldots, n\}$, the derivatives of the coordinates $\{\dot{x}_i = dx_i/dt, i = 1, \ldots, n\}$, and the line integral J_n may all depend on a parameter α that parameterizes the path of integration. The parameterized line integral may be written as

$$J_n(\alpha) = \int_{t_1}^{t_2} f_n(x_1(t, \alpha), \dot{x}_1(t, \alpha), \ldots, x_n(t, \alpha), \dot{x}_n(t, \alpha), t) dt \tag{14.3.1}$$

and the condition for finding the path of integration that minimizes $J_n(\alpha)$ is

$$\frac{\partial J_n(\alpha)}{\partial \alpha}\bigg|_{\alpha=0} = 0 \tag{14.3.2}$$

The procedure for deriving the Euler–Lagrange equations for several dependent variables is analogous to the procedure used in Section 14.1 for a single dependent variable.

We derive the Euler–Lagrange equations for several dependent variables by first applying Eq. (14.3.2) to Eq. (14.3.1) to obtain

$$\frac{\partial J_n}{\partial \alpha}\bigg|_{\alpha=0} = \int_{t_1}^{t_2} \sum_{i=1}^{n} \left(\frac{\partial f_n}{\partial x_i} \frac{\partial x_i}{\partial \alpha} + \frac{\partial f_n}{\partial \dot{x}_i} \frac{\partial \dot{x}_i}{\partial \alpha} \right) dt = 0 \tag{14.3.3}$$

Integrating the second term in parentheses by parts gives

$$\begin{aligned}
\int_{t_1}^{t_2} \frac{\partial f_n}{\partial \dot{x}_i} \frac{\partial \dot{x}_i}{\partial \alpha} dt &= \int_{t_1}^{t_2} \frac{\partial f_n}{\partial \dot{x}_i} \frac{\partial^2 x_i}{\partial t \partial \alpha} dt \\
&= \frac{\partial f_n}{\partial \dot{x}_i} \frac{\partial x_i}{\partial \alpha}\bigg|_{t_1}^{t_2} - \int_{t_1}^{t_2} \frac{d}{dt}\left(\frac{\partial f_n}{\partial \dot{x}_i} \right) \frac{\partial x_i}{\partial \alpha} dt
\end{aligned} \tag{14.3.4}$$

for each value of i. We observe that the partial derivative $\partial x_i/\partial \alpha$ is arbitrary except at the end points $\{t_1, t_2\}$, where $\partial x_i/\partial \alpha$ must equal zero to guarantee that all of the parameterized paths of integration pass through the end points for any value of the parameter α. Consequently, the first term on the right-hand side of Eq. (14.3.4) vanishes and Eq. (14.3.4) simplifies to

$$\int_{t_1}^{t_2} \frac{\partial f_n}{\partial \dot{x}_i} \frac{\partial \dot{x}_i}{\partial \alpha} dt = - \int_{t_1}^{t_2} \frac{d}{dt}\left(\frac{\partial f_n}{\partial \dot{x}_i} \right) \frac{\partial x_i}{\partial \alpha} dt \tag{14.3.5}$$

Substituting Eq. (14.3.5) into Eq. (14.3.3) gives

$$\frac{\partial J_n}{\partial \alpha}\bigg|_{\alpha=0} = \int_{t_1}^{t_2} \sum_{i=1}^{n} \left[\left(\frac{\partial f_n}{\partial x_i} - \frac{d}{dt}\frac{\partial f_n}{\partial \dot{x}_i} \right) \frac{\partial x_i}{\partial \alpha} \right] dt = 0 \tag{14.3.6}$$

The extremum condition $(\partial J_n(\alpha)/\partial \alpha)|_{\alpha=0} = 0$ is satisfied when the integrand in Eq. (14.3.6) vanishes at $\alpha = 0$. The integrand vanishes for the set of arbitrary deformations $\{\partial x_i/\partial \alpha\}$ when the term in parentheses vanishes; therefore the extremum condition $(\partial J_n(\alpha)/\partial \alpha)|_{\alpha=0} = 0$ is satisfied when

$$\frac{\partial f_n}{\partial x_i} - \frac{d}{dt}\left(\frac{\partial f_n}{\partial \dot{x}_i} \right) = 0 \text{ for all } i = 1, \ldots, n \tag{14.3.7}$$

The equations represented by Eq. (14.3.7) are Euler–Lagrange equations for n dependent variables.

EXERCISE 14.5: Solve Eq. (14.3.7) for the function of two dependent variables $f_2(x_1, x_2, \dot{x}_1, \dot{x}_2, t) = (m/2)(\dot{x}_1^2 + \dot{x}_2^2) - (k/2)(x_1 + x_2)^2$ and $\{m, k\}$ are constants.

14.4 CALCULUS OF VARIATIONS WITH CONSTRAINTS

Suppose the line integral J_n in Eq. (14.3.1) is subject to the m constraints

$$g_j(x_1(t, \alpha), \dot{x}_1(t, \alpha), \ldots, x_n(t, \alpha), \dot{x}_n(t, \alpha), t) = 0, j = 1, \ldots, m \qquad (14.4.1)$$

The constraints can be included in the calculus of variations using the method of undetermined multipliers.

Let's assume that the constraints in Eq. (14.4.1) are differentiable so that

$$\frac{\partial g_j}{\partial \alpha} = \sum_{i=1}^{n} \frac{\partial g_j}{\partial x_i} \frac{\partial x_i}{\partial \alpha} = 0, j = 1, \ldots, m \qquad (14.4.2)$$

We include the constraints in the calculus of variations by multiplying $\partial g_j / \partial \alpha$ by the unspecified multiplier λ_j and then integrate with respect to t; thus

$$\int_{t_1}^{t_2} \lambda_j \sum_{i=1}^{n} \frac{\partial g_j}{\partial x_i} \frac{\partial x_i}{\partial \alpha} \, dt = \int_{t_1}^{t_2} \sum_{i=1}^{n} \lambda_j \frac{\partial g_j}{\partial x_i} \frac{\partial x_i}{\partial \alpha} \, dt = 0, j = 1, \ldots, m \qquad (14.4.3)$$

Each of the m constraint equations in Eq. (14.4.3) is now added to Eq. (14.3.6) to give

$$\left. \frac{\partial J_n}{\partial \alpha} \right|_{\alpha=0} = \int_{t_1}^{t_2} \sum_{i=1}^{n} \left[\left(\frac{\partial f_n}{\partial x_i} - \frac{d}{dt} \frac{\partial f_n}{\partial \dot{x}_i} + \sum_{j=1}^{m} \left(\lambda_j \frac{\partial g_j}{\partial x_i} \right) \right) \frac{\partial x_i}{\partial \alpha} \right] dt = 0 \qquad (14.4.4)$$

The n derivatives $\{\partial x_i / \partial \alpha\}$ in Eq. (14.4.4) are not arbitrary because of the constraint equations in Eq. (14.4.2). We recognize this dependence by rewriting Eq. (14.4.4) as

$$\int_{t_1}^{t_2} \sum_{i=1}^{n-m} \left[\left(\frac{\partial f_n}{\partial x_i} - \frac{d}{dt} \frac{\partial f_n}{\partial \dot{x}_i} + \sum_{j=1}^{m} \left(\lambda_j \frac{\partial g_j}{\partial x_i} \right) \right) \frac{\partial x_i}{\partial \alpha} \right] dt$$

$$+ \int_{t_1}^{t_2} \sum_{i=n-m+1}^{n} \left[\left(\frac{\partial f_n}{\partial x_i} - \frac{d}{dt} \frac{\partial f_n}{\partial \dot{x}_i} + \sum_{j=1}^{m} \left(\lambda_j \frac{\partial g_j}{\partial x_i} \right) \right) \frac{\partial x_i}{\partial \alpha} \right] dt = 0$$

$$(14.4.5)$$

We choose the m undetermined multipliers $\{\lambda_1, \ldots, \lambda_m\}$ so that the parenthetic term in the second integral of Eq. (14.4.5) vanishes. The result is

$$\frac{\partial f_n}{\partial x_i} - \frac{d}{dt}\left(\frac{\partial f_n}{\partial \dot{x}_i}\right) + \sum_{j=1}^{m} \lambda_j \frac{\partial g_j}{\partial x_i} = 0 \text{ for all } i = n - m + 1, \ldots, n \qquad (14.4.6)$$

With this choice, the remaining $n-m$ derivatives $\{\partial x_1/\partial \alpha, \ldots, \partial x_{n-m}/\partial \alpha\}$ are arbitrary. Equation (14.4.5) is satisfied by recognizing that the second integral vanishes because of our choice of undetermined multipliers in Eq. (14.4.6), and by requiring that the parenthetic term in the first integral must vanish for arbitrary $\{\partial x_1/\partial \alpha, \ldots, \partial x_{n-m}/\partial \alpha\}$ so that

$$\frac{\partial f_n}{\partial x_i} - \frac{d}{dt}\left(\frac{\partial f_n}{\partial \dot{x}_i}\right) + \sum_{j=1}^{m} \lambda_j \frac{\partial g_j}{\partial x_i} = 0 \text{ for all } i = 1, \ldots, n - m \qquad (14.4.7)$$

Equations (14.4.2), (14.4.6), and (14.4.7) are $n + m$ equations for the n variables $\{x_1, \ldots, x_n\}$ and the m undetermined multipliers $\{\lambda_1, \ldots, \lambda_m\}$. Equations (14.4.6) and (14.4.7) are Euler–Lagrange equations with constraints.

EXERCISE 14.6: Find the Euler–Lagrange equations with constraints for the function of three dependent variables $f_3(r, \theta, z) = (m/2)[\dot{r}^2 + r^2\dot{\theta}^2 + \dot{z}^2] - mgz$ subject to the constraint $r^2 = cz$, where $\{m, g, c\}$ are constants. Note that f_3 is expressed in cylindrical coordinates $\{r, \theta, z\}$.

CHAPTER 15

TENSOR ANALYSIS

Tensor analysis is the study of tensors and their transformation properties. A few important physical quantities that may be classified as tensors include temperature, velocity, stress, permeability, and the electromagnetic tensor that contains both the electric and magnetic fields. The purpose of this chapter is to review essential concepts from tensor analysis. For additional discussion, see such references as Arfken and Weber [2001, Sec. 2.6], Chow [2000, Chap. 1], Zwillinger [2003, Sec. 5.10], Lebedev and Cloud [2003], Boas [2005], and Eddington [1960, Chap. II].

15.1 CONTRAVARIANT AND COVARIANT VECTORS

The concept of a vector is extended in tensor analysis to account for the different ways the components of a vector are expressed when the coordinate representation of a vector is transformed from one coordinate system to another. In this section we use two transformation relations to motivate the definition of two types of vectors: contravariant vectors and covariant vectors.

Contravariant Vector

Consider the representation of a three-dimensional vector in two coordinate systems: an unprimed coordinate system with components $\{x^i\} = \{x^1,\ x^2,\ x^3\}$ and a primed coordinate system with components $\{x'^i\} = \{x'^1,\ x'^2,\ x'^3\}$. We

Math Refresher for Scientists and Engineers, Third Edition By John R. Fanchi
Copyright © 2006 John Wiley & Sons, Inc.

write the component index as a superscript in conformity with the modern notation for a contravariant vector. The definition of contravariant vector depends on the transformation properties of the vector. In general, the primed components of a three-dimensional vector are related to the unprimed components by the coordinate transformation

$$x'^1 = f^1(x^1, x^2, x^3)$$
$$x'^2 = f^2(x^1, x^2, x^3)$$
$$x'^3 = f^3(x^1, x^2, x^3)$$

$$(15.1.1)$$

where the functions $\{f^1, f^2, f^3\}$ relate the components in the unprimed coordinate system to the components in the primed coordinate system. The definition of a contravariant vector is motivated by considering the transformation properties of the displacement vector.

Example: Suppose the primed coordinate system is obtained by rotating the unprimed coordinate system counterclockwise by an angle θ about the x^3 axis. From Section 4.1, we know that the components in the primed coordinate system are related to the components in the unprimed coordinate system by

$$x'^1 = x^1 \cos\theta + x^2 \sin\theta$$
$$x'^2 = -x^1 \sin\theta + x^2 \cos\theta$$
$$x'^3 = x^3$$

$$(15.1.2)$$

The relationships in Eq. (15.1.2) are the transformation equations that connect the unprimed and primed coordinate systems.

The components of the displacement vector in the unprimed coordinate system $\{x^i\}$ are the differentials $\{dx^i\} = \{dx^1, dx^2, dx^3\}$. The components of the displacement vector in the primed coordinate system are expressed in terms of the components of the displacement vector in the unprimed coordinate system by

$$dx'^i = \frac{\partial x'^i}{\partial x^1} dx^1 + \frac{\partial x'^i}{\partial x^2} dx^2 + \frac{\partial x'^i}{\partial x^3} dx^3, \ i = 1, 2, 3$$

or

$$dx'^i = \sum_{j=1}^{3} \frac{\partial x'^i}{\partial x^j} dx^j, i = 1, 2, 3$$

$$(15.1.3)$$

The partial derivatives $\partial x'^i / \partial x^j$ are determined from the transformation relations given by Eq. (15.1.1). Equation (15.1.3) is the transformation equation for the displacement vector. The generalization of Eq. (15.1.3) to n dimensions is used as the definition of a set of vectors known as contravariant vectors. A *contravariant vector* $\{A^i\}$ in n-dimensional space is a vector with n components that transforms according to the relation·

$$A'^i = \sum_{j=1}^{n} \frac{\partial x'^i}{\partial x^j} A^j, \ i = 1, 2, \ldots, n \tag{15.1.4}$$

EXERCISE 15.1: Compare the set of equations in Eq. (15.1.1) with the set of equations in Eq. (15.1.2). What are the functions $\{f^1, f^2, f^3\}$? Use these functions in Eq. (15.1.3) to calculate the components of the displacement vector in the primed (rotated) coordinate system.

Covariant Vector

The transformation equation for contravariant vectors given by Eq. (15.1.4) is one way the components of a vector may transform. Another transformation equation is obtained by considering the transformation properties of a vector $\{b_i\}$ that is obtained as the gradient of a function $\phi(x^1, x^2, x^3)$ in three-dimensional space. Notice that the component index is written as a subscript to distinguish the type of vector that we are now considering from a contravariant vector. Vector $\{b_i\}$ is known as a covariant vector. Also notice that the function ϕ depends on the components of a contravariant vector in the unprimed coordinate system. The components of $\{b_i\}$ are the components of the gradient of ϕ, namely,

$$b_i = \frac{\partial \phi}{\partial x^i}, \ i = 1, 2, 3 \tag{15.1.5}$$

The components of the vector $\{b'_i\}$ in the primed coordinate system are

$$b'_i = \frac{\partial \phi}{\partial x'^i} = \sum_{j=1}^{3} \frac{\partial \phi}{\partial x^j} \frac{\partial x^j}{\partial x'^i}, \ i = 1, 2, 3 \tag{15.1.6}$$

Equation (15.1.6) is the transformation equation for the gradient of a function $\phi(x^1, x^2, x^3)$. It may be rewritten using Eq. (15.1.5) to obtain

$$b'_i = \sum_{j=1}^{3} \frac{\partial x^j}{\partial x'^i} b_j, \ i = 1, 2, 3 \tag{15.1.7}$$

Equation (15.1.7) suggests a generalization analogous to Eq. (15.1.4). The generalization of Eq. (15.1.7) to n dimensions defines a set of vectors called covariant vectors. A *covariant vector* $\{B_i\}$ in n-dimensional space is a vector with n components that transforms according to the relation

$$B'_i = \sum_{j=1}^{n} \frac{\partial x^j}{\partial x'^i} B_j, \ i = 1, 2, \ldots, n \tag{15.1.8}$$

Application: Galilean and Lorentz Transformations. The Galilean transformation is a set of rules for transforming the physical description of a moving object in one uniformly moving (nonaccelerating) coordinate system to another. Each coordinate system is referred to as a reference frame. A simple expression for the Galilean transformation can be obtained by considering the relative motion of the two reference frames shown in Figure 15.1. The primed reference frame F' is moving with a constant speed v in the x direction relative to the unprimed reference frame F. Using Cartesian coordinates, the origins of the two reference frames F and F' are at $x' = x = 0$ at time $t' = t = 0$. The Galilean transformation is

$$
\begin{aligned}
x' &= x - vt \\
y' &= y \\
z' &= z \\
t' &= t
\end{aligned}
\qquad (15.1.9)
$$

The spatial coordinate x' depends on the values of x and t in the unprimed reference frame, while the time coordinates t and t' are equal. The equality of t and t' expresses Isaac Newton's view of time as absolute and immutable.

The Galilean transformation does not change the form of Newton's laws, but it does change the form of Maxwell's equations. The change in the form of Maxwell's equations following a Galilean transformation implied that the speed of light should depend on the motion of the observer. Measurements by Michelsen and Morley showed that the speed of light does not depend on the motion of observers whose relative motion is uniform. Einstein proposed a set of transformation rules that preserved the form of Maxwell's equations in reference frames that were moving uniformly with respect to each other. These transformation rules, known collectively as the Lorentz transformation, introduced a mutual dependence between space and time coordinates in the primed and unprimed reference frames. The relationship between the unprimed and primed space and time coordinates is given by the

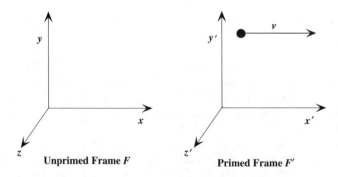

Figure 15.1 Two reference frames in uniform relative motion.

Lorentz transformation

$$x' = \frac{x - vt}{\sqrt{1 - (v/c)^2}} = \gamma x - \gamma vt; \; \gamma = \frac{1}{\sqrt{1 - (v/c)^2}}$$

$$y' = y$$
$$z' = z$$ (15.1.10)

$$t' = \frac{t - x(v/c^2)}{\sqrt{1 - (v/c)^2}} = \gamma t - \gamma \frac{v}{c^2} x$$

where $\{v, c, \gamma\}$ are constants with respect to the space and time coordinates. Equation (15.1.10) shows that space and time depend on one another in both the primed and unprimed reference frames. Equations (15.1.9) and (15.1.10) are examples of linear coordinate transformations in four-dimensional space.

EXERCISE 15.2: Show that the set of equations in Eq. (15.1.9) is a special case of Eq. (15.1.10). *Hint*: Apply the limit $c \to \infty$ to the set of equations in Eq. (15.1.10).

15.2 TENSORS

A *tensor* in an n-dimensional space is a function whose value at a given point in space does not depend on the coordinate system used to cover the space. Examples of tensors include temperature, velocity, and stress. Temperature T is a scalar tensor, or tensor of rank zero, because it is defined by one number and does not depend on a component index. The contravariant vector $\{A^i\}$ and covariant vector $\{B_j\}$ are tensors of rank one; they depend on one component index. Tensors with higher ranks are obtained by generalizing the definitions of contravariant and covariant vectors. General definitions for the transformation equations of second-rank tensors and for tensors of arbitrary rank are presented in this section.

Summation Convention and Tensor Notation

The summation convention says that a sum is implied over any index that appears twice in a term. For example, the transformation equation in Eq. (15.1.4) for a contravariant vector $\{A^i\}$ is

$$A'^i = \frac{\partial x'^i}{\partial x^j} A^j$$ (15.2.1)

and the transformation equation in Eq. (15.1.8) for a covariant vector $\{B_j\}$ is

$$B'_i = \frac{\partial x^j}{\partial x'^i} B_j$$ (15.2.2)

Each index $\{i, j\}$ in n-dimensional space is an integer that is understood to range from 1 to n. The summation convention simplifies the notation, but may cause confusion if the use of the summation convention is not clearly stated.

We have used curly brackets $\{A^i\}$ and $\{B_j\}$ to indicate a tensor. We now simplify the notation by omitting the brackets. In this widely used notation, the contravariant vector $\{A^i\}$, which is a tensor of rank one, is written as A^i. The number of indices indicates the rank of the tensor, and the placement of the index as a superscript or subscript indicates the transformation property of the tensor. A superscript index indicates the contravariant transformation property illustrated in Eq. (15.2.1) for a contravariant tensor of rank one, and a subscript index indicates the covariant transformation property illustrated in Eq. (15.2.2) for a covariant tensor of rank one.

Second-Rank Tensors

A tensor of rank two, or a second-rank tensor, in an n-dimensional space may be formed by combining first-rank tensors. For example, the product of the first-rank tensors A^i and B_j is the tensor C^i_j with n^2 components. The ij^{th} component is

$$C^i_j = A^i B_j \qquad (15.2.3)$$

The tensor C^i_j is called a *mixed tensor* because it has both an upper (contravariant) index i and a lower (covariant) index j. Equation (15.2.3) defines the ij^{th} component of a second-rank tensor that is formed by the *direct product* of two first-rank tensors. The direct product of two first-rank tensors has also been called the *outer product*. The *inner product* of two first-rank tensors is the scalar

$$C = A^i B_i \qquad (15.2.4)$$

Note that the summation convention is implied in Eq. (15.2.4).

Two other second-rank tensors may be formed by the direct product of two first-rank tensors. The outer product of two contravariant tensors A^i and D^j is the contravariant second-rank tensor E^{ij} with the n^2 components

$$E^{ij} = A^i D^j \qquad (15.2.5)$$

The outer product of two covariant tensors B_i and F_j is the covariant second-rank tensor G_{ij} with the n^2 components

$$G_{ij} = B_i F_j \qquad (15.2.6)$$

The second-rank tensors C^i_j, E^{ij}, G_{ij} have been formed by the direct product of two first-rank tensors. Second-rank tensors do not have to be formed as the outer product of first-rank tensors. In general, the transformation equation for a

TABLE 15.1   Transformation Equations for Second-Rank Tensors

Second-Rank Tensor	Transformation Equation
Contravariant	$A'^{ij} = \dfrac{\partial x'^i}{\partial x^a} \dfrac{\partial x'^j}{\partial x^b} A^{ab}$
Covariant	$A'_{ij} = \dfrac{\partial x^a}{\partial x'^i} \dfrac{\partial x^b}{\partial x'^j} A_{ab}$
Mixed	$A'^i_j = \dfrac{\partial x'^i}{\partial x^a} \dfrac{\partial x^b}{\partial x'^j} A^a_b$

contravariant second-rank tensor is

$$A'^{ij} = \frac{\partial x'^i}{\partial x^a} \frac{\partial x'^j}{\partial x^b} A^{ab} \tag{15.2.7}$$

where we have used the summation convention. In this case, the two indices $\{a, b\}$ appear twice in the term on the right-hand side of Eq. (15.2.7). A summation is implied for each index.

A second-rank tensor is a quantity that has $n \times n$ components and satisfies one of the transformation equations in Table 15.1. Each transformation equation in Table 15.1 represents $n \times n$ equations for the $n \times n$ components of the second-rank tensor.

EXERCISE 15.3: Write the transformation equations in Table 15.1 without the summation convention; that is, explicitly show the summations for each equation.

Tensors with Arbitrary Rank

The transformation equation for a tensor with arbitrary rank is

$$A'^{a_1 a_2 \ldots a_n}_{b_1 b_2 \ldots b_m} = \frac{\partial x'^{a_1}}{\partial x^{c_1}} \frac{\partial x'^{a_2}}{\partial x^{c_2}} \cdots \frac{\partial x'^{a_n}}{\partial x^{c_n}} \frac{\partial x^{d_1}}{\partial x'^{b_1}} \frac{\partial x^{d_2}}{\partial x'^{b_2}} \cdots \frac{\partial x^{d_m}}{\partial x'^{b_m}} A^{c_1 c_2 \ldots c_n}_{d_1 d_2 \ldots d_m} \tag{15.2.8}$$

where the summation convention has been used. The tensor in Eq. (15.2.8) has n contravariant indices and m covariant indices. Sums are implied over the $n + m$ repeated indices $\{c_1, c_2, \ldots, c_n, d_1, d_2, \ldots, d_m\}$ on the right-hand side of Eq. (15.2.8).

EXERCISE 15.4: Use Eq. (15.2.8) to write the transformation equation for a tensor with two contravariant indices and a single covariant index.

15.3 THE METRIC TENSOR

The length of a vector with components $\{x^i, i = 1, 2, 3\}$ in a three-dimensional Euclidean space can be written in several equivalent ways; thus

$$
x \cdot x = \vec{x} \cdot \vec{x} = (x^1)^2 + (x^2)^2 + (x^3)^2 = \sum_{j=1}^{3} x^j x_j = \sum_{i=1}^{3} \sum_{j=1}^{3} g_{ij} x^i x^j,
\tag{15.3.1}
$$

$$
x^j = x_j
$$

The elements of the matrix $\{g_{ij}\}$ are

$$
\{g_{ij}\} = \begin{bmatrix} 1 & 0 & 0 \\ 0 & 1 & 0 \\ 0 & 0 & 1 \end{bmatrix}
\tag{15.3.2}
$$

The matrix $\{g_{ij}\}$ is called the *metric tensor* or the *fundamental tensor* because it determines the length of a vector in a particular geometric space. The tensor g_{ij} is a *second-rank tensor*. The determinant of the matrix $\{g_{ij}\}$ representing the metric tensor must be nonzero. In our example, the tensor g_{ij} is defined for a three-dimensional space and is written as a 3×3 matrix in Eq. (15.3.2). Equation (15.3.2) illustrates the observation that a second-rank tensor in n-dimensional space can be written as a square matrix with $n \times n$ elements.

The elements of the metric tensor depend on the coordinate system we are working in. In the case of the metric tensor for Euclidean space shown in Eq. (15.3.2), there is no difference between the contravariant vector x^j and the covariant vector x_j. This is not true in some geometric spaces, such as Riemann space.

The relation

$$
g^{ik} g_{kj} = \delta^i_j
\tag{15.3.3}
$$

defines the contravariant tensor g^{ij} in terms of the covariant tensor g_{ij} and the Kronecker delta function

$$
\delta^i_j = \begin{cases} 1 \text{ if } i = j \\ 0 \text{ if } i \neq j \end{cases}
\tag{15.3.4}
$$

The Kronecker delta function δ^i_j is a mixed tensor of rank two and δ^i_j is the same in every coordinate system.

If we write the matrix $\{g_{ij}\}$ as $\mathbf{g}$ and its inverse as $\mathbf{g}^{-1}$, the inverse $\mathbf{g}^{-1}$ is obtained from the relation

$$
\mathbf{g}\mathbf{g}^{-1} = \mathbf{g}^{-1}\mathbf{g} = \mathbf{I}
\tag{15.3.5}
$$

where $\mathbf{I}$ is the identity matrix. Equation (15.3.5) is equivalent to Eq. (15.3.3) and shows that the contravariant tensor g^{ij} is equal to $\mathbf{g}^{-1}$, the inverse of matrix $\mathbf{g}$.

Example: The inverse of matrix $\mathbf{g}$ in three-dimensional Euclidean space is found from the matrix equation

$$\mathbf{g}\,\mathbf{g}^{-1} = \mathbf{g}^{-1}\mathbf{g} = \begin{bmatrix} 1 & 0 & 0 \\ 0 & 1 & 0 \\ 0 & 0 & 1 \end{bmatrix}$$

and gives the result

$$\mathbf{g}^{-1} = \mathbf{g} = \begin{bmatrix} 1 & 0 & 0 \\ 0 & 1 & 0 \\ 0 & 0 & 1 \end{bmatrix}$$

The contravariant tensor g^{ij} is equal to the covariant tensor g_{ij} in this case; thus

$$\{g^{ij}\} = \{g_{ij}\} = \begin{bmatrix} 1 & 0 & 0 \\ 0 & 1 & 0 \\ 0 & 0 & 1 \end{bmatrix}$$

EXERCISE 15.5: Write Eq. (15.3.1) using the summation convention.

EXERCISE 15.6: Show that the determinant of the metric tensor represented by the matrix in Eq. (15.3.2) is nonzero.

Transformation of the Metric Tensor

The transformation equation for the metric tensor g_{ij} is obtained from Table 15.1:

$$g'_{ij} = \frac{\partial x^a}{\partial x'^i} \frac{\partial x^b}{\partial x'^j} g_{ab} \tag{15.3.6}$$

Application: Energy–Momentum Four-Vector. The geometry of the special theory of relativity is defined in a four-dimensional space. The four components of the four-dimensional space are a time coordinate and three space coordinates. A widely used notation expresses the indices for the components of a space–time four-vector in the range 0 to 3. The 0th component of the space–time four-vector is the time coordinate $x^0 = ct$, where c is the speed of light in vacuum, and the three remaining components $\{x^1, x^2, x^3\}$ represent the three space coordinates. The four-dimensional space of the special theory of relativity is called *Minkowski space–time*.

The length s of a four-vector with space–time components $\{x^0, x^1, x^2, x^3\}$ in Minkowski space–time is

$$s^2 = \sum_{\mu=0}^{3} \sum_{\nu=0}^{3} g_{\mu\nu} x^\mu x^\nu \text{ for } \mu, \nu = 0, 1, 2, 3 \tag{15.3.7}$$

The fundamental metric for determining the length s of a four-vector in Minkowski space–time is

$$[g_{\mu\nu}] = \begin{bmatrix} 1 & 0 & 0 & 0 \\ 0 & -1 & 0 & 0 \\ 0 & 0 & -1 & 0 \\ 0 & 0 & 0 & -1 \end{bmatrix} \tag{15.3.8}$$

Substituting the metric of Minkowski space–time into Eq. (15.3.7) gives

$$s^2 = (x^0)^2 - (x^1)^2 - (x^2)^2 - (x^3)^2 = c^2 t^2 - (x^1)^2 - (x^2)^2 - (x^3)^2 \tag{15.3.9}$$

The length s of the space–time four-vector is a scalar. The four-vector is called timelike if $s^2 > 0$ and it is called spacelike if $s^2 < 0$. If an object is moving at the speed of light, then the space–time four-vector has length $s^2 = 0$.

Another four-vector that has special importance is the energy–momentum four-vector $\{p^\mu\} = \{p^0, p^1, p^2, p^3\} = \{E/c, p^1, p^2, p^3\}$. The 0th component of the energy–momentum four-vector is the total energy E divided by the speed of light in vacuum. The remaining three spatial components are the components of the momentum vector. The length of the energy–momentum four-vector for a relativistic object with mass m is given by Eq. (15.3.7) with the space–time four-vector replaced by the energy–momentum four-vector; thus

$$m^2 c^2 = (p^0)^2 - (p^1)^2 - (p^2)^2 - (p^3)^2 = \frac{E^2}{c^2} - (p^1)^2 - (p^2)^2 - (p^3)^2 \tag{15.3.10}$$

The energy–momentum four-vector is timelike if $m^2 c^2 > 0$ and it is spacelike if $m^2 c^2 < 0$. An object moving slower than the speed of light is called a bradyon and has a positive mass. It is mathematically possible for an object to be moving faster than the speed of light. Faster-than-light objects are called tachyons. In the context of Eq. (15.3.10), the mass of a faster-than-light object would have to be imaginary. There are other ways to treat tachyons, but that discussion is beyond the scope of this book [e.g., see Fanchi, 1993]. If an object is moving at the speed of light, then the energy–momentum four-vector has length $m^2 c^2 = 0$.

EXERCISE 15.7: Show that Eq. (15.3.10) reduces to the famous result $E \approx mc^2$ when the momentum of the mass is small compared to the mass of the object.

15.4 TENSOR PROPERTIES

Tensors have many properties that facilitate their use in applications. Some tensor properties are summarized in this section.

Equality of Tensors

A tensor $A_{d_1 d_2 \ldots d_m}^{c_1 c_2 \ldots c_n}$ with n contravariant indices and m covariant indices has the *contravariant rank* n and the *covariant rank* m. Two tensors **A**, **B** are equal if they have the same contravariant rank, the same covariant rank, and every component satisfies the equality $A_{d_1 d_2 \ldots d_m}^{c_1 c_2 \ldots c_n} = B_{d_1 d_2 \ldots d_m}^{c_1 c_2 \ldots c_n}$.

Contraction

The operation of contraction can be applied to a mixed tensor. A mixed tensor has both a contravariant index and a covariant index. For example, the tensor $A_{b_1 b_2 \ldots b_j \ldots b_m}^{a_1 a_2 \ldots a_i \ldots a_n}$ is a mixed tensor if the contravariant rank $n > 0$ and the covariant rank $m > 0$. Contraction is achieved by setting a contravariant index (such as a_i) equal to a covariant index (such as b_j) and summing over the index. The contraction operation reduces the rank of a tensor by two; the contravariant rank is reduced from n to $n - 1$, and the covariant rank is reduced from m to $m - 1$.

Example: The contraction of a mixed second-rank tensor A_j^i is given by A_i^i, where the summation convention is implied. The contracted second-rank tensor $A = A_i^i$ is a tensor of rank zero (the scalar A).

Addition and Subtraction of Tensors

Tensors with the same contravariant rank and the same covariant rank may be added to and subtracted from each other by adding and subtracting their components. Let $A_{d_1 d_2 \ldots d_m}^{c_1 c_2 \ldots c_n}, B_{d_1 d_2 \ldots d_m}^{c_1 c_2 \ldots c_n}, C_{d_1 d_2 \ldots d_m}^{c_1 c_2 \ldots c_n}$ represent tensors that have contravariant rank n and covariant rank m. The addition and subtraction of tensors is given by

$$A_{d_1 d_2 \ldots d_m}^{c_1 c_2 \ldots c_n} \pm B_{d_1 d_2 \ldots d_m}^{c_1 c_2 \ldots c_n} = C_{d_1 d_2 \ldots d_m}^{c_1 c_2 \ldots c_n} \tag{15.4.1}$$

Example: The sum of two contravariant second-rank tensors is found from Eq. (15.4.1) to be $A^{ij} \pm B^{ij} = C^{ij}$.

Products of Tensors

The *outer product* of two tensors is a tensor whose contravariant rank is the sum of the contravariant ranks of the two tensors, and whose covariant rank is the sum of the covariant rank of the two tensors. Suppose $A_{b_1 b_2 \ldots b_m}^{a_1 a_2 \ldots a_n}$ is a tensor with contravariant rank n and covariant rank m, and $B_{d_1 d_2 \ldots d_q}^{c_1 c_2 \ldots c_p}$ is a tensor with contravariant rank p and covariant rank q. The outer product forms a tensor with contravariant rank

$n + p$ and covariant rank $m + q$. The components of the new tensor are

$$A^{a_1 a_2 \ldots a_n}_{b_1 b_2 \ldots b_m} B^{c_1 c_2 \ldots c_p}_{d_1 d_2 \ldots d_q} = C^{a_1 a_2 \ldots a_n, c_1 c_2 \ldots c_p}_{b_1 b_2 \ldots b_m, d_1 d_2 \ldots d_q} \tag{15.4.2}$$

As noted previously, the outer product of two tensors is also known as the direct product.

The *inner product* of two tensors $A^{a_1 a_2 \ldots a_n}_{b_1 b_2 \ldots b_m}$, $B^{c_1 c_2 \ldots c_p}_{d_1 d_2 \ldots d_q}$ is obtained by contracting the outer product of the two tensors with respect to one contravariant index and one covariant index.

Example: The outer product of two vectors is a second-rank tensor that is sometimes called the *tensor product*. If the vectors are the first-rank tensors $\mathbf{a}, \mathbf{b}$, the tensor product may be written as $\mathbf{a} \otimes \mathbf{b}$. The components of the tensor product of two first-rank contravariant tensors are given by $a^i b^j = c^{ij}$. The tensor $\mathbf{c}$ is a second-rank contravariant tensor. The tensor product is used to define a mathematical operation called the *wedge product* [Goldstein et al., 2002, Sec. 7.5]. The wedge product is defined as $\mathbf{a} \wedge \mathbf{b} = \mathbf{a} \otimes \mathbf{b} - \mathbf{b} \otimes \mathbf{a}$ with components given by $(\mathbf{a} \wedge \mathbf{b})^{ij} = a^i b^j - b^i a^j$.

Example: The inner product of a contravariant vector $\mathbf{a}$ and a covariant vector $\mathbf{b}$ is $a^i b_i = c$. The tensor $\mathbf{c}$ is a tensor of rank 0; that is, $\mathbf{c}$ is a scalar. Alternatively, we can form the outer product $a^i b_j = c^i_j$ and then form the contraction $c^i_i = c$, where the summation convention is implied.

Symmetry and Antisymmetry

A contravariant tensor $A^{a_1 \ldots a_i \ldots a_j \ldots a_n}$ with rank $n \geq 2$ is symmetric if, for all values of two contravariant indices a_i, a_j, we have the equality $A^{a_1 \ldots a_i \ldots a_j \ldots a_n} = A^{a_1 \ldots a_j \ldots a_i \ldots a_n}$ when the order of the indices a_i, a_j has been interchanged. The contravariant tensor $B^{a_1 \ldots a_i \ldots a_j \ldots a_n}$ with rank $n \geq 2$ is antisymmetric if we have the equality $B^{a_1 \ldots a_i \ldots a_j \ldots a_n} = -B^{a_1 \ldots a_j \ldots a_i \ldots a_n}$ when the indices a_i, a_j are interchanged. If a tensor is antisymmetric, the components with $a_i = a_j$ must be zero since $B^{a_1 \ldots a_i \ldots a_i \ldots a_n} = -B^{a_1 \ldots a_i \ldots a_i \ldots a_n} = 0$. Note that we are *not* using the summation convention in this case. Analogous definitions apply to covariant tensors. The definitions of symmetry and antisymmetry apply to two contravariant indices or to two covariant indices; they do not apply when one index is contravariant and the other index is covariant.

Example: The metric tensor g_{ij} is symmetric; that is, $g_{ij} = g_{ji}$.

Example: A second-rank contravariant tensor C^{ij} is symmetric if $C^{ij} = C^{ji}$. It is antisymmetric if $C^{ij} = -C^{ji}$. The symmetric and antisymmetric properties can be used to resolve the second-rank tensor C^{ij} into symmetric and antisymmetric parts when we express the second-rank tensor C^{ij} as the identity

$$C^{ij} = \frac{1}{2}(C^{ij} + C^{ji}) + \frac{1}{2}(C^{ij} - C^{ji}) \tag{15.4.3}$$

The first term on the right-hand side of Eq. (15.4.3) is a symmetric tensor, and the second term is an antisymmetric tensor.

Associated Tensors

An associated tensor is formed by taking the inner product of a tensor with the metric tensor g_{ij} or g^{ij}. Suppose $A^{a_1 a_2 \dots j \dots a_n}_{b_1 b_2 \dots b_m}$ is a tensor with contravariant rank n and covariant rank m. The inner product of $A^{a_1 a_2 \dots j \dots a_n}_{b_1 b_2 \dots b_m}$ with the covariant metric tensor g_{ij} gives the associated tensor

$$A^{a_1 a_2 \dots a_{n-1}}_{b_1 b_2 \dots i \dots b_{m+1}} = g_{ij} A^{a_1 a_2 \dots j \dots a_n}_{b_1 b_2 \dots b_m} \tag{15.4.4}$$

The inner product with g_{ij} lowers the contravariant index j and introduces another covariant index i.

If we use the contravariant metric tensor g^{ij} to form the inner product with $A^{a_1 a_2 \dots a_n}_{b_1 b_2 \dots j \dots b_m}$, we obtain a new associated tensor

$$A^{a_1 a_2 \dots i \dots a_{n+1}}_{b_1 b_2 \dots b_{m-1}} = g^{ij} A^{a_1 a_2 \dots a_n}_{b_1 b_2 \dots j \dots b_m} \tag{15.4.5}$$

The inner product with g^{ij} raises the covariant index j and introduces another contravariant index i.

Example: The tensor associated with the contravariant vector A^j is the covariant vector $A_i = g_{ij} A^j$. Conversely, the tensor associated with the covariant vector A_j is the contravariant vector $A^i = g^{ij} A_j$. The summation convention is implied in both cases.

Line Element

The scalar line element ds is determined from the metric tensor g_{ij} and the displacement vector dx^i as

$$ds^2 = dx_j \, dx^j = g_{ij} \, dx^i dx^j \tag{15.4.6}$$

where the summation convention is implied.

Christoffel Symbols

The metric tensor is used to form the *Christoffel symbol of the first kind*

$$[ij, k] = \frac{1}{2} \left\{ \frac{\partial g_{ik}}{\partial x^j} + \frac{\partial g_{jk}}{\partial x^i} - \frac{\partial g_{ij}}{\partial x^k} \right\} \tag{15.4.7}$$

and the *Christoffel symbol of the second kind*

$$\{ij, k\} = g^{kl}[ij, l] = \frac{1}{2} g^{kl} \left\{ \frac{\partial g_{il}}{\partial x^j} + \frac{\partial g_{jl}}{\partial x^i} - \frac{\partial g_{ij}}{\partial x^l} \right\} \tag{15.4.8}$$

The shape of the bracket in Eqs. (15.4.7) and (15.4.8) distinguishes the kind of Christoffel symbol. Christoffel symbols look like third-rank tensors, but Christoffel symbols are *not* tensors. The Christoffel symbols vanish when the elements of the metric tensor are constant, which is the case in Cartesian coordinates. Christoffel symbols satisfy the symmetry relations

$$[ij, k] = [ji, k] \; \{ij, k\} = \{ji, k\} \tag{15.4.9}$$

Covariant Derivatives

Differentiation in tensor calculus is defined so that the mathematical quantity obtained by differentiating a tensor also transforms as a tensor. The mathematical quantity formed by the differentiation process is a tensor that is called the *covariant derivative*. Covariant derivatives of vectors and second-rank tensors are given in Table 15.2. The semicolon in the notation $A_{\cdots;j}$ indicates covariant differentiation with respect to x^j. The covariant derivative of a tensor of rank n includes the usual partial derivative term plus n terms that depend on Christoffel symbols of the second kind $\{ij, k\}$. The terms with a Christoffel symbol as a factor are needed to assure that the covariant derivative transforms as a tensor.

EXERCISE 15.8: Calculate the contraction of the Kronecker delta function δ_j^i in n dimensions.

TABLE 15.2   Covariant Derivatives of Vectors and Second-Rank Tensors

Tensor	Covariant Derivative
Contravariant vector A^i	$A^i_{;j} = \dfrac{\partial A^i}{\partial x^j} + \{kj, i\} A^k$
Covariant vector A_i	$A_{i;j} = \dfrac{\partial A_i}{\partial x^j} - \{ij, k\} A_k$
Contravariant tensor A^{ij}	$A^{ij}_{;k} = \dfrac{\partial A^{ij}}{\partial x^k} + \{lk, i\} A^{lj} + \{lk, j\} A^{il}$
Covariant tensor A_{ij}	$A_{ij;k} = \dfrac{\partial A_{ij}}{\partial x^k} - \{ik, l\} A_{lj} - \{jk, l\} A_{il}$
Mixed tensor A_i^j	$A^j_{i;k} = \dfrac{\partial A^j_i}{\partial x^k} - \{ik, l\} A^j_l + \{lk, j\} A^l_i$

EXERCISE 15.9: Calculate the line element for the metric tensors

$$[g_{ij}] = \begin{bmatrix} 1 & 0 \\ 0 & -1 \end{bmatrix} \quad \text{and} \quad [g_{ij}] = \begin{bmatrix} -\left(1 - \dfrac{2m}{x^1}\right)^{-1} & 0 \\ 0 & 1 - \dfrac{2m}{x^1} \end{bmatrix}$$

Assume the parameter m is a constant and x^1 is component 1 in a two-dimensional space with the displacement vector dx^i.

EXERCISE 15.10: Show that $[ij, k] + [kj, i] = \partial g_{ik} / \partial x^j$, where $[\cdots]$ is the Christoffel symbol of the first kind.

CHAPTER 16

PROBABILITY

Probability and statistics play a vital role in modern analysis. The concepts of probability and statistics are based on set theory, which is briefly reviewed in Section 16.1 as a prelude to our discussion of probability theory in the remainder of this chapter, probability distributions in Chapter 17, and descriptive statistics in Chapter 18. Further information can be found in a variety of references, such as Walpole and Myers [1985], Larsen and Marx [1985], Kurtz [1991], Blaisdell [1998], Miller and Miller [1999], and Kreyszig [1999].

16.1 SET THEORY

A set is a collection of points, or objects, which satisfy specified criteria. For example, if we flip a two-sided coin N tosses, and each toss outcome is either a head H or a tail T, each outcome is an element of the set. If the set of N elements represents all possible outcomes of the experiment, then the set S is called the sample space S. There is only one element in the set S corresponding to each outcome of the experiment, such as the toss of a coin. An element in the sample space is called a sample point or a simple event. An event is a subset of the sample space S. A discrete sample space contains a finite number of elements, while a continuous sample space typically contains the outcomes of measurements of continuous physical properties such as speed and temperature.

Math Refresher for Scientists and Engineers, Third Edition By John R. Fanchi
Copyright © 2006 John Wiley & Sons, Inc.

The intersection of two sets A, B is denoted as $A \cap B$ and defines those elements that are in both sets. If two sets do not have any elements in common, the events are said to be mutually exclusive and the sets are said to be disjoint. The union of two sets is written $A \cup B$. It consists of those elements that belong to set A, or to set B, or to both sets. The collection of elements in the sample space S comprises a set called the *universal set U*. The universal set is the union of a set A with its complement A' {read "not A"), thus $U = A \cup A'$. The *empty set $\emptyset$* is a set with no elements. It is the intersection of A and A', thus $\emptyset = A \cap A'$. The empty set is a subset of every set, which we write as $\emptyset \subset A$.

Suppose A, B, C are sets of objects in some sample space S. The union of sets $\{A, B, C\}$ comprise the universal set U covering the sample space S; that is, $A \cup B \cup C = U$. The sets A, B, C are said to belong to U, or are subsets of U, which is written as $A, B, C \subset U$. Sets satisfy the following properties:

PROPERTIES OF SETS

Closure	For each pair of sets A, B there exists a unique set $A \cup B$ and unique set $A \cap B$ in S
Commutative	$A \cup B = B \cup A$ and $A \cap B = B \cap A$
Associative	$(A \cup B) \cup C = A \cup (B \cup C)$
	$(A \cap B) \cap C = A \cap (B \cap C)$
Distributive	$A \cap (B \cup C) = (A \cap B) \cup (A \cap C)$
	$A \cup (B \cap C) = (A \cup B) \cap (A \cup C)$

A graphical tool for studying sets is the Venn diagram. The sample space is represented by a rectangle while the subsets are represented by circles within the rectangle. An example of a Venn diagram is shown in Chapter 1.

Permutations

If k objects are selected from n distinct objects, any particular arrangement of these k objects is called a *permutation*. In general, the total number of permutations of n objects taken k at a time is

$$P_{n,k} = \frac{n!}{(n-k)!}$$

When the number of arranged objects k is equal to the number of distinct objects n, the number of permutations of n objects taken n at a time is

$$P_{n,n} = n!$$

Example: Let the sample space S contain three objects so that $S = \{A, B, C\}$. The possible permutations are shown as follows:

PERMUTATIONS OF THREE OBJECTS

Taken 1 at a Time	Taken 2 at a Time	Taken 3 at a Time
A, B, C	AB, AC, BA, BC, CB, CA	$ABC, ACB, BAC, BCA,$ CAB, CBA
$P_{3,1} = 3!/(3 - 1)! = 3$	$P_{3,2} = 3!/(3 - 2)! = 6$	$P_{3,3} = 3!/(3 - 3)! = 6$

Notice that the order of the objects A, B, C matters; that is, ABC and CBA are two separate permutations. Also observe that there is no duplication, or replacement, of any of the objects; for example, AAC is not an acceptable permutation.

Combinations

If k objects are selected from n distinct objects, any particular group of k objects is called a *combination*. The number of combinations of n things taken k at a time equals the number of permutations divided by $k!$. The primary difference between permutation and combination is the importance of order. Order must be considered in calculating permutations, but order is not important when calculating combinations. Thus the familiar "combination" lock found in many American schools is actually a permutation lock because the order of the numbers has significance.

Example: The number of permutations of three things $\{a\,b\,c\}$ taken three at a time is calculated by counting each possible sequence—thus $\{a\,b\,c\}$, $\{a\,c\,b\}$, $\{c\,a\,b\}$, $\{c\,b\,a\}$, $\{b\,a\,c\}$, and $\{b\,c\,a\}$. There are $3! = 6$ permutations of three things $\{a\,b\,c\}$ taken three at a time. If the order of the three things in the sequence does not matter, then there is only one combination of the three things $\{a\,b\,c\}$.

The number of combinations of n things taken k at a time is

$$C_{n,k} = \frac{P_{n,k}}{k!} = \frac{n!}{k!(n-k)!}$$

The symbol $\binom{n}{k} = C_{n,k}$ is often used to represent the number of combinations of n things taken k at a time. The symbol $\binom{n}{k}$ is called the *binomial coefficient* because it appears in the binomial series

$$(x + y)^n = \sum_{k=0}^{n} \binom{n}{k} x^{n-k} y^k$$

where n is a positive integer.

EXERCISE 16.1: Calculate $P_{5,4}$, $C_{5,2}$, $C_{5,3}$, $C_{6,6}$, and $C_{6,0}$.

16.2   PROBABILITY DEFINED

Probability may be defined as either an objective or a subjective probability. An objective probability is a probability that can be calculated using a repeatable, well-defined procedure. For example, the number of times that an event can occur in a set of possible outcomes is the frequency of occurrence of the event, and can be used as an estimate of probability. Thus if an experiment has n different, equally probable outcomes, and if m of these correspond to event A, then the probability of event A is $P(A) = m/n$.

Subjective probability is not as well defined as objective probability. Subjective probability is an estimate of probability based on prior knowledge, experience, or simple guessing. Subjective probabilities are justified when very little direct evidence is available; otherwise it is preferable to work with objective probabilities.

Probability Tree Diagrams

The preparation of a tree diagram is a systematic method for analyzing a sequence of events in which branching occurs. Tree diagrams make it possible to explicitly portray and then count all of the possible sequences of events. They can be used to calculate objective probabilities.

Example: A restaurant offers a choice of four different salads $\{S_i : 1 \leq i \leq 4\}$, three different main dishes $\{M_j : 1 \leq j \leq 3\}$, and two desserts $\{D_k : 1 \leq k \leq 2\}$. How many distinct meals are offered?

Figure 16.1 displays each sequence $\{S_i\, M_j\, D_k\}$, where the indices have the ranges $\{1 \leq i \leq 4,\ 1 \leq j \leq 3,\ 1 \leq k \leq 2\}$. We must sum each sequence to determine the number of distinct meals. The final number is 24, or $4 \times 3 \times 2$ meals.

EXERCISE 16.2: Four fair coins are each flipped once. Use a probability tree diagram to determine the probability that at least two coins will show heads.

16.3   PROPERTIES OF PROBABILITY

Probability is expected to satisfy a few basic postulates. Let A be an event in the sample space S, and let ϕ be the empty set. Then probability P has the following

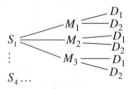

Figure 16.1   A tree diagram.

fundamental properties:

$$P(A) \geq 0$$
$$P(S) = 1$$
$$P(\phi) = 0$$

The first postulate requires that probability is a nonnegative number. The second postulate guarantees that at least one possibility in the sample space S will occur. The third postulate says that the probability of obtaining the empty set is zero. If events A and A' are complementary, then $P(A) + P(A') = 1$.

The probability that two events A and B will both occur is given by the additive rule $P(A \cup B) = P(A) + P(B) - P(A \cap B)$. The term $P(A \cap B)$ is zero if events A and B are mutually exclusive or disjoint. The additive rule can be extended to three or more events. In the case of three events A, B, and C, the additive rule is

$$P(A \cup B \cup C) = P(A) + P(B) + P(C) - P(A \cap B) - P(A \cap C)$$
$$- P(B \cap C) + P(A \cap B \cap C)$$

Conditional and Marginal Probabilities

The probability that event A occurred when it is known that event B occurred is called the *conditional probability* of A given B. The conditional probability of event A given event B is symbolized as $P(A|B)$. If A and B are two subsets of a sample space S, then the conditional probability of A relative to B is

$$P(A|B) = \frac{P(A \cap B)}{P(B)}$$

if $P(B) > 0$. The probability $P(B)$ is called the *marginal probability* of event B. Multiplying the above equation by $P(B)$ gives the multiplicative rule $P(A \cap B) = P(A|B)P(B)$.

If A and B are any two subsets of a sample space S, then the multiplicative rule can be expressed in two symmetric forms: $P(A \cap B) = P(B)P(A|B)$ and $P(A \cap B) = P(A)P(B|A)$. The multiplicative rule for three events is $P(A \cap B \cap C) = P(A)P(B|A)P(C|A \cap B)$.

Example: Find the probability of drawing three aces from a deck of cards without replacement. Event A denotes an ace obtained on the first draw, event B denotes an ace obtained on the second draw, and event C denotes an ace obtained on the third draw.

The probability of drawing an ace on the first draw from a 52-card deck with 4 aces is $P(A) = 4/52$. The probability of drawing an ace on the second draw given that an ace has already been drawn and not replaced is $P(B|A) = 3/51$.

The probability of drawing an ace on the third draw given that aces have already been drawn and not replaced on the first two draws is $P(C|A \cap B) = 2/50$. The probability of drawing three aces without replacement is $P(A \cap B \cap C) = P(A) P(B|A)P(C|A \cap B) = (4/52)(3/51)(2/50)$.

Example: Suppose one hundred people qualify for teaching positions. The distribution of backgrounds is given in the following table and represents the sample space for this problem. A review of the table shows that some of the people are married, and some are single; some have three or more years of experience, and some have less than three years experience. Suppose we must select an applicant at random from this sample space. Use the multiplicative rule $P(A \cap B) = P(B)P(A|B)$ to determine the probability that the applicant will have three or more years experience given that the applicant is married.

Distribution of Backgrounds	Married (M)	Single (M')	Total
Three or more years experience (E)	12	24	36
Less than three years experience (E')	18	46	64
Total	30	70	100

Let E be the selection of an applicant with three or more years experience, and M be the selection of an applicant who is married. The marginal probability that the applicant has three or more years experience is $P(E) = 36/100$. The marginal probability that the applicant is married is $P(M) = 30/100$. The joint probability that the applicant is married and has three or more years experience is $P(E \cap M) = 12/100$. The conditional probability that the applicant will have three or more years experience given that the applicant is married is $P(E|M) = P(E \cap M)/P(M) = 12/30$.

Independent Events

Two events A and B are said to be *statistically independent* if the probability of occurrence of event A is not affected by the occurrence or nonoccurrence of event B, and vice versa. In terms of conditional probabilities, two events A and B are statistically independent if $P(A|B) = P(A)$ and $P(B|A) = P(B)$. The multiplication rule for two independent events is $P(A \cap B) = P(A)P(B)$. The general multiplication rule for N independent events is

$$P(A_1 \cap A_1 \cap \cdots \cap A_N) = \prod_{i=1}^{N} P(A_i)$$

Random Variables

Suppose we are given a sample space S. We can map elements of S onto a real number axis using a function X whose domain is S and whose range is the set of real numbers that maps elements of S onto a real number axis. The function X is called a *random variable*. A random variable X is a discrete random variable if its set of possible outcomes is a finite or countable number of values on the real number axis. The random variable X is a continuous random variable if it requires a continuum of values on the real number axis.

Bayes' Theorem

Bayes' theorem relates conditional and unconditional probabilities for events A and B through the multiplicative rule $P(A \cap B) = P(A)P(B|A)$. The joint probability $P(A \cap B)$ is the probability that both events A and B will occur. It is the product of the probability that event B will occur given that event A has occurred and the probability that event A will occur. Thus the occurrence of event A gives us information about the likelihood that event B will occur. If we recall the symmetric forms of the multiplicative rule, namely, $P(A \cap B) = P(B)P(A|B)$ and $P(A \cap B) = P(A)P(B|A)$, we can derive the equality $P(B)P(A|B) = P(A)P(B|A)$ or

$$P(A|B) = \frac{P(B|A)P(A)}{P(B)}$$

This expression provides a model for learning from observations by updating information with new evidence. From this perspective, the updated probability $P(A|B)$ is proportional to the initial or *prior* probability $P(A)$ that event A will occur times the likelihood $P(B|A)$ of observing event B given event A. The probability $P(B)$ normalizes the updated or *posterior* probability $P(A|B)$. The *prior* probability $P(A)$ refers to a probability that does not account for new information, whereas the *posterior* probability $P(A|B)$ refers to a probability that does account for new information.

EXERCISE 16.3: Two unbiased six-sided dice, one red and one green, are tossed and the number of dots appearing on their upper faces are observed. The results of a dice roll can be expressed as the ordered pair (g, r), where g denotes the green die and r denotes the red die.

(a) What is the sample space S?
(b) What is the probability of throwing a seven?
(c) What is the probability of throwing a seven or a ten?
(d) What is the probability that the red die shows a number less than or equal to three and the green die shows a number greater than or equal to five?
(e) What is the probability that the green die shows a one, given that the sum of the numbers on the two dice is less than four?

2	3	2	1
1	2	2	3
3	3	1	3
2	3	3	2

Figure 16.2 Dart board.

Application: Dart Board. The meaning of conditional probabilities is illustrated using a simple example from Collins [1985]. Consider a "dart board" with area elements numbered and shaded as shown in Figure 16.2. There are 16 elements, all equally likely to be hit by the toss of a dart. The probability of the joint event of a number j and a color c is

$$P(j,c) = \frac{N(j,c)}{N}$$

where $N(j,c)$ is the number of squares showing the number j and the color c while N is the total number of squares.

The unconditional, or marginal, probability for a color c is

$$P'(c) = \sum_j \frac{N(j,c)}{N} = \frac{N(c)}{N}$$

where $N(c)$ is the total number of squares showing the color c without regard for the number contained in the square. Then the conditional probability for j to occur, given that the color c has been observed, is

$$P_c(j|c) = \frac{P(j,c)}{P'(c)} = \frac{N(j,c)}{N(c)}$$

from the definitions given above. Conditional probability applies to a reduction in the range of possible events. For example, if c is "dark," D, then the conditional probability $P_c(j|c)$ refers to a reduced dart board as in Figure 16.3. Specific numerical results for this example are $P(2,D) = 1/16$ and $P_c(2|D) = 1/6$.

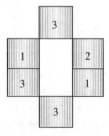

Figure 16.3 Reduced dart board.

16.4   PROBABILITY DISTRIBUTION DEFINED

The concept of random variable is used here as the independent variable associated with probability distributions. The type of probability distribution depends on the system of interest and the associated random variable. For example, the most obvious distinction is between discrete and continuous systems. The probability distribution of a discrete random variable is discrete, and the probability distribution of a continuous random variable is continuous. Discrete and continuous probability distributions are discussed here.

Discrete Probability Distribution

Let us define a function $f(x)$ as the probability of a random variable X associated with the outcome x; thus $f(x) = P(X = x)$. The ordered pair $\{x, f(x)\}$ is called the *probability distribution* or *probability function* of the discrete random variable X if

$$\text{(a)} \quad f(x) \geq 0$$
$$\text{(b)} \quad \sum_x f(x) = 1$$

If we wish to calculate the probability that the outcome associated with a random variable X is less than or equal to the real number x, then we must work with the cumulative distribution of the random variable X. In the case of a discrete random variable X, the cumulative distribution $F(x)$ with probability distribution $f(x)$ is

$$F(x) = P(X \leq x) = \sum_i f(x_i), \quad -\infty < x < \infty,$$

where the summation extends over those values of i such that $x_i \leq x$.

Continuous Probability Distribution

Continuous probability distributions are defined in terms of a probability density function $\rho(X)$ for a continuous random variable X defined over a set of real numbers x. The probability density function, which is often called the *probability density*, must be positive and its integral with respect to x over the entire sample space must be unity. The mathematical expressions for these requirements are $\rho(x) \geq 0$ and the normalization condition

$$\int_{-\infty}^{\infty} \rho(x)\,dx = 1$$

for all values of x in the interval $-\infty < x < \infty$.

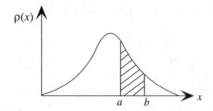

Figure 16.4 Cumulative distribution $P(a < X < b)$.

The cumulative distribution function for a continuous random variable X with probability density $\rho(x)$ is

$$F(x) = P(X \leq x) = \int_{-\infty}^{x} \rho(t)\, dt, \quad -\infty < x < \infty$$

where t is a dummy variable of integration. If the cumulative distribution $F(x)$ is known, the probability density can be found from the derivative

$$\rho(x) = \frac{dF(x)}{dx}$$

Using the above definitions and relationships lets us write the cumulative distribution in the form

$$P(a < X < b) = F(b) - F(a) = \int_{a}^{b} \rho(t)\, dt$$

The probability $P(a < X < b)$ is the area under the curve shown in Figure 16.4.

CHAPTER 17

PROBABILITY DISTRIBUTIONS

Fundamental concepts of probability theory were presented in Chapter 16 along with a list of references. Joint, conditional, and marginal probability distributions are defined here, followed by the presentation of examples of probability distributions and the parameters used to characterize them.

17.1 JOINT PROBABILITY DISTRIBUTION

In many cases the behavior of a system must be described by two or more random variables. The probability distribution associated with two or more random variables is called a *joint probability distribution*. It is illustrated below for the case of two random variables X and Y. The extension of these ideas to k random variables is given in Section 17.3.

Discrete Case

If the random variables X and Y are discrete, a function $f(x, y)$ of the real numbers x, y is a joint probability distribution if $f(x, y) \geq 0$ for all values of x and y, and the sum of $f(x, y)$ over all values of x and y must be unity. The latter requirement

Math Refresher for Scientists and Engineers, Third Edition By John R. Fanchi
Copyright © 2006 John Wiley & Sons, Inc.

is the normalization condition

$$\sum_x \sum_y f(x, y) = 1$$

If these conditions are satisfied, the probability of $X = x$ and $Y = y$ is $P(X = x, Y = y) = f(x, y)$.

An important distribution known as the marginal distribution can be obtained from the joint distribution function. In particular, the marginal distributions of discrete random variables X and Y are

$$g(x) = \sum_y f(x, y) \text{ and } h(y) = \sum_x f(x, y)$$

respectively. Marginal distributions provide information about a single random variable selected from a distribution with more than one random variable.

Conditional distributions can be obtained from joint and marginal distributions by dividing the joint distribution by the marginal distribution. For example, the conditional distribution of the random variable Y given $X = x$ is

$$f(y|x) = \frac{f(x, y)}{g(x)} \text{ if } g(x) \geq 0$$

A similar relationship holds for $f(x|y)$; thus

$$f(x|y) = \frac{f(x, y)}{h(y)} \text{ if } h(y) \geq 0$$

Rearranging the expressions for conditional probability gives

$$f(x, y) = g(x)f(y|x) = h(y)f(x|y)$$

The random variables X and Y are statistically independent if the joint distribution equals the product of the marginal distributions, or $f(x, y) = g(x) h(y)$. Statistical independence implies that the conditional probability of a random variable is equal to the marginal probability of the random variable. In our case, the random variables X and Y are statistically independent if $f(x|y) = g(x)$ and $f(y|x) = h(x)$.

Continuous Case

If the random variables X and Y are continuous, a joint probability density function $p(x, y)$ can be defined if the function $p(x, y)$ satisfies the nonnegative requirement $p(x, y) \geq 0$ for all values of (x, y) and the normalization condition

$$\int_{-\infty}^{\infty} \int_{-\infty}^{\infty} p(x, y) \, dx \, dy = 1$$

The probability of obtaining the random variables X and Y in the region R of the $x-y$ plane is

$$P[(X, Y) \in R] = F(x, y) = \iint_R \rho(x, y) \, dx \, dy$$

The joint probability density can be obtained from the joint probability distribution function $F(x, y)$ by calculating

$$\rho(x, y) = \frac{\partial^2 F(x, y)}{\partial x \, \partial y}$$

The marginal distributions for continuous random variables X and Y are

$$\rho(x) = \int_{-\infty}^{\infty} \rho(x, y) \, dy \text{ and } \rho(y) = \int_{-\infty}^{\infty} \rho(x, y) \, dx$$

Conditional probability distributions for the continuous random variables X and Y are given by

$$\rho(x|y) = \frac{\rho(x, y)}{\rho(y)} \text{ and } \rho(y|x) = \frac{\rho(x, y)}{\rho(x)}$$

Rearranging these expressions gives the joint probability in the form

$$\rho(x, y) = \rho(x|y) \, \rho(y) = \rho(y|x) \, \rho(x)$$

As in the discrete case, statistical independence corresponds to the condition that $\rho(x, y) = \rho(x) \, \rho(y)$, or $\rho(x|y) = \rho(x)$ and $\rho(y|x) = \rho(y)$.

Example: Suppose the joint probability distribution function $F(x, y) = (1 - e^{-x}) (1 - e^{-y})$ for $x, y > 0$. $F(x, y)$ has the properties $F(0, 0) = 0$ and $F(\infty, \infty) = 1$. The probability density is

$$\rho(x, y) = \frac{\partial^2 F(x, y)}{\partial x \, \partial y} = e^{-x} e^{-y}$$

EXERCISE 17.1: What is the probability that x, y are in the intervals $1 \le x \le 2$ and $1 \le y \le 3$ for the joint probability distribution function $F(x, y) = (1 - e^{-x}) (1 - e^{-y})$ for $x, y > 0$. $F(x, y)$ has the properties $F(0, 0) = 0$ and $F(\infty, \infty) = 1$, and $\rho(x, y) = e^{-x} e^{-y}$.

Example: Suppose we have a joint probability density $\rho(x, y) = 2(x + 2y)/3$ that is valid in the intervals $0 \le x \le 1$ and $0 \le y \le 1$. The conditional probability density $\rho(x|y)$ is obtained by first calculating the marginal distribution function

$$\rho(y) = \int\limits_0^1 \tfrac{2}{3} \, (x + 2y) \, dx = \tfrac{2}{3} \left(\tfrac{1}{2} + 2y \right)$$

and then using the definition of conditional probability to obtain

$$\rho(x|y) = \frac{\rho(x, y)}{\rho(y)} = \frac{(x + 2y)}{\left(\tfrac{1}{2} + 2y \right)}$$

EXERCISE 17.2: Use the conditional probability $\rho(x|y) = (x + 2y)/(\tfrac{1}{2} + 2y)$ defined in the intervals $0 \le x \le 1$ and $0 \le y \le 1$ to determine the probability that x will have a value less than $\tfrac{1}{2}$, given $y = \tfrac{1}{2}$.

Example: The statistical independence of the variables x and y for the joint probability density $\rho(x, y) = 2x^2 y^2/81$ is determined by calculating the marginal probability densities. The variables x and y are defined in the interval $0 < x < y < 3$. The marginal density for x is

$$\rho(x) = \int\limits_x^3 \frac{2x^2 y^2}{81} \, dy = \frac{2}{81} x^2 \left(9 - \frac{x^3}{3} \right)$$

and the marginal density for y is

$$\rho(y) = \int\limits_0^y \frac{2x^2 y^2}{81} \, dx = \frac{2}{81} y^2 \left(\frac{y^3}{3} \right)$$

Taking the product of marginal densities gives

$$\rho(x) \, \rho(y) = \left(\frac{2}{81} \right)^2 x^2 y^2 \left(9 - \frac{x^3}{3} \right) \left(\frac{y^3}{3} \right) \ne \rho(x, y)$$

which demonstrates that x and y are statistically dependent. Notice that the definition of the interval established a dependence of the random variables.

EXERCISE 17.3: Determine the statistical independence of the variables x and y for the joint probability density $\rho(x, y) = x^2 y^2/81$ by calculating the marginal probability densities. The variables x and y are defined in the intervals $0 < x < 3$ and $0 < y < 3$. Note the difference between this exercise and the interval defined in the preceding example.

17.2   EXPECTATION VALUES AND MOMENTS

Probability distributions are characterized by expectation values and moments. These quantities are defined below.

Expectation Values

Much useful information about the expected behavior of a particular system can be obtained from a probability distribution. The quantities used to extract this information are referred to as *expectation values*. The average value of a set of measurements is an example of an expectation value. Let X be a random variable with probability density $f(x)$. Then the expected value of $X, E(X)$, is defined as

$$E(X) = \sum_x xf(x)$$

if X is a discrete random variable, and

$$E(X) = \int_{-\infty}^{\infty} xf(x)\, dx$$

if X is a continuous random variable. The expected value $E(X)$ is usually referred to as the *average value*. In general, the expected value of a function g of a random variable X is defined as

$$E\left[g(X)\right] = \sum_x g(x) \cdot f(x)$$

if X is a discrete random variable and

$$E\left[g(X)\right] = \int_{-\infty}^{\infty} g(x) \cdot \rho(x)\, dx$$

if X is a continuous random variable.

Given the above definitions of expectation value in terms of sums and integrals, it is straightforward to derive the following useful properties:

1. $E\left[aX + bY\right] = aE(X) + bE(Y)$.
2. $E\left[X \cdot Y\right] = E(X) \cdot E(Y)$ if X and Y are statistically independent.

Moments

Moments About the Origin   The moments about the origin of a probability distribution are the expected values of the random variable that has the given distribution. The rth moment of X, usually denoted by μ'_r, is defined as

$$\mu'_r = E[X^r] = \sum_x x^r f(x)$$

if X is a discrete random variable and

$$\mu'_r = E[X^r] = \int_{-\infty}^{\infty} x^r f(x) \, dx$$

if X is a continuous random variable. The first moment, μ'_r, is called the *mean* of random variable X and is usually denoted by μ.

EXERCISE 17.4: Calculate the first moment, or mean, of the discrete distribution

$$f(x) = \binom{N}{x} p^x (1 - p)^{N-x}$$

with the binomical coefficient $\binom{N}{x}$ for $0 \leq x \leq N$, and p a constant.

Moments About the Mean   The rth moment about the mean, usually denoted by μ_r, is defined as

$$\mu_r = E[(X - \mu)^r] = \sum_x (x - \mu)^r f(x)$$

if X is a discrete random variable and

$$\mu_r = E[(X - \mu)^r] = \int_{-\infty}^{\infty} (x - \mu)^r f(x) \, dx$$

if X is a continuous random variable.

The second moment about the mean, μ_2, is

$$\mu_2 = E[(X - \mu)^2] = \mu'_2 - \mu^2$$

The moment μ_2 is called the *variance* of the random variable X and is denoted by σ^2. The square root of the variance σ is called the *standard deviation*.

Moment Generating Functions   The moment generating function of the random variable X is defined in terms of the expected value $E(e^{tX})$ of the exponential function e^{tX}, where t is a dummy parameter. In particular, the moment generating function is

$$m_x(t) = E(e^{tX}) = \sum_x e^{tx}f(x)$$

if X is a discrete random variable and

$$m_x(t) = E(e^{tX}) = \int_{-\infty}^{\infty} e^{tx}f(x)\, dx$$

if X is a continuous random variable. The rth moment about the origin is

$$\mu'_r = \left.\frac{d^r m_x(t)}{dt^r}\right|_{t=0}$$

for $r = 0, 1, 2, \ldots$.

The series expansion

$$m_x(t) = E(e^{tX}) = E\left[1 + Xt + \frac{(Xt)^2}{2!} + \cdots\right] = 1 + \mu'_1\, t + \mu'_2\frac{t^2}{2} + \cdots$$

shows that the moments about the origin μ'_r appear as coefficients of $t^r/r!$. The function $m_x(t)$ may be regarded as generating the moments. The moments about the mean μ_r may be generated by the generating function

$$M_x(t) = E[e^{t(X-\mu)}] = e^{-\mu t}E(e^{tX}) = e^{-\mu t}\, m_x(t)$$

Example: The moment generating function of the normal distribution (described in Section 17.4) is

$$m_x(t) = \exp\left(\mu t + \frac{\sigma^2 t^2}{2}\right)$$

The mean is the first moment given by

$$\left.\frac{d}{dt}m_x(t)\right|_{t=0} = \left.\frac{d}{dt}\exp\left(\mu t + \frac{\sigma^2 t^2}{2}\right)\right|_{t=0}$$

$$= (\mu + \sigma^2 t)\exp\left(\mu t + \frac{\sigma^2 t^2}{2}\right)\Big|_{t=0} = \mu$$

17.3 MULTIVARIATE DISTRIBUTIONS

Much of the discussion presented above for one or two random variables can readily be extended to distributions with k random variables. Distributions with more than one random variable are called *multivariate distributions*. Properties of multivariate distributions are presented here.

Discrete Case

The k-dimensional random variable $(X_1, X_2, \ldots, X_k)$ is a k-dimensional discrete random variable if it assumes values only at a finite or denumerable number of points $(x_1, x_2, \ldots, x_k)$. The joint probability $f(x_1, x_2, \ldots, x_k)$ of the k-dimensional random variable is defined as

$$P[X_1 = x_1, X_2 = x_2, \ldots, X_k = x_k] = f(x_1, x_2, \ldots, x_k)$$

for every value that the set of random variables can assume. If E is a subset of the set of values that the random variables can assume, then the probability of E is

$$P(E) = P[(X_1, X_2, \ldots, X_k) \text{ is in } E] = \sum_E f(x_1, x_2, \ldots, x_k)$$

where the sum is over all those points in E. The cumulative distribution is defined as

$$F(x_1, x_2, \ldots, x_k) = \sum_{x_1} \sum_{x_2} \cdots \sum_{x_k} f(x_1, x_2, \ldots, x_k)$$

Continuous Case

The k random variables $X_1, X_2, \ldots, X_k$ are said to be jointly distributed if there exists a function ρ such that $\rho(x_1, x_2, \ldots, x_k) \geq 0$ for all $-\infty < x_i < \infty$, $i = 1, 2, \ldots, k$ and the probability of an event E in the sample space is given by

$$P(E) = P[(X_1, X_2, \ldots, X_k) \text{ is in } E] = \int_E \rho(x_1, x_2, \ldots, x_k) dx_1 \, dx_2 \cdots dx_k$$

The function $\rho(x_1, x_2, \ldots, x_k)$ is called the *joint probability density* of the random variables $X_1, X_2, \ldots, X_k$.

The cumulative distribution is defined as

$$F(x_1, x_2, \ldots, x_k) = \int_{-\infty}^{x_1} \int_{-\infty}^{x_2} \cdots \int_{-\infty}^{x_k} \rho(x_1, x_2, \ldots, x_k) \, dx_k \cdots dx_2 \, dx_1$$

Given the cumulative distribution, the joint probability density can be obtained by forming the derivative

$$\rho(x_1, x_2, \ldots, x_k) = \frac{\partial}{\partial x_1} \times \frac{\partial}{\partial x_2} \times \cdots \times \frac{\partial}{\partial x_k} F(x_1, x_2, \ldots, x_k)$$

Expectation Values and Moments

The rth moment of the ith random variable X_i is defined in terms of expectation values as

$$E(X_i^r) = \sum_{x_1} \sum_{x_2} \cdots \sum_{x_k} x_i^r f(x_1, x_2, \ldots, x_k)$$

if the X_i are discrete random variables and

$$E(X_i^r) = \int_{-\infty}^{\infty} \int_{-\infty}^{\infty} \cdots \int_{-\infty}^{\infty} x_i^r \rho\,(x_1, x_2, \ldots, x_k)\, dx_1\ dx_2 \cdots dx_k$$

if the X_i are continuous random variables. Joint moments about the origin are defined as the expectation value

$$E(X_1^{r_1} X_2^{r_2} \cdots X_k^{r_k})$$

where $r_1 + r_2 + \cdots + r_k$ is the order of the moment. Joint moments about the mean are defined as

$$E[(X_1 - \mu_1)^{r_1}(X_2 - \mu_2)^{r_2} \cdots (X_k - \mu_k)^{r_k}]$$

Joint moment generating functions for k random variables may also be defined. The reader should consult the references for further details.

Marginal and Conditional Distributions

If the k discrete random variables $X_1, X_2, \ldots, X_k$ have the joint distribution $f(x_1, x_2, \ldots, x_k)$, then the marginal distribution of the subset of random variables $\{X_i : i = 1, \ldots, p \text{ with } p < k\}$ is

$$g(x_1, x_2, \ldots, x_p) = \sum_{x_{p+1}} \sum_{x_{p+2}} \cdots \sum_{x_k} f(x_1, x_2, \ldots, x_k)$$

The conditional distribution of the subset of discrete random variables $X_1, X_2, \ldots, X_p$ given $X_{p+1}, X_{p+2}, \ldots, X_k$ is

$$h(x_1, x_2, \ldots, x_p | x_{p+1}, x_{p+2}, \ldots, x_k) = \frac{f(x_1, x_2, \ldots, x_k)}{g(x_{p+1}, x_{p+2}, \ldots, x_k)}$$

$$\text{if } g(x_{p+1}, x_{p+2}, \ldots, x_k) \geq 0$$

Similarly, if the k random variables $X_1, X_2, \ldots, X_k$ are continuous with a joint density $\rho(x_1, x_2, \ldots, x_k)$, then the marginal distribution of $\{X_i : i = 1, \ldots, p$ with $p < k\}$ is

$$\rho(x_1, x_2, \ldots, x_p) = \int_{-\infty}^{\infty} \int_{-\infty}^{\infty} \cdots \int_{-\infty}^{\infty} \rho(x_1, x_2, \ldots, x_k) \, dx_{p+1} \cdots dx_{k-1} \, dx_k$$

The conditional distribution of the subset of continuous random variables $X_1, X_2, \ldots, X_p$ given $X_{p+1}, X_{p+2}, \ldots, X_k$ is

$$\rho(x_1, x_2, \ldots, x_p | x_{p+1}, x_{p+2}, \ldots, x_k) = \frac{\rho(x_1, x_2, \ldots, x_k)}{\rho(x_{p+1}, x_{p+2}, \ldots, x_k)}$$

$$\text{if } \rho(x_{p+1}, x_{p+2}, \ldots, x_k) \geq 0$$

Variance and Covariance

The variance σ_i^2 of a random variable X_i is defined as the expectation value

$$\sigma_i^2 = E[(X_i - \mu_i)^2]$$

The square root of the variance σ_i is the standard deviation. The covariance σ_{ij} of random variables X_i and X_j is given by the expectation value

$$\sigma_{ij} = r_{ij} \sigma_i \sigma_j = E[(X_i - \mu_i)(X_j - \mu_j)]$$

where r_{ij} is the correlation coefficient and σ_i and σ_j are the standard deviations of X_i and X_j. The variance is equal to the covariance σ_{ii} in which $i = j$ and the corresponding correlation coefficient is $r_{ii} = 1$.

EXERCISE 17.5: Suppose we are given the joint probability density $\rho(x, y, z) = (x + y)e^{-z}$ for three random variables defined in the intervals $0 < x < 1$, $0 < y < 1$, and $0 < z$.

(a) Find the joint probability density $\rho(x, y)$.
(b) Find the marginal probability density $\rho(y)$.

Application: Quantum Mechanics and Probability Theory. Max Born first introduced the idea that quantum mechanics and probability were linked in 1926. Since then, the physics community has debated the precise meaning of the quantum mechanical wavefunction. One interpretation relies on a direct relationship between quantum theory and probability theory. The relationship is based on a derivation of quantum mechanical wave equations from fundamental assumptions in probability. An example of this derivation is outlined here.

Construction of the nonrelativistic Schrödinger equation from fundamental probability concepts begins with the assumption that a conditional probability density $\rho(y|t)$ exists. The symbol y denotes a set of three spatial coordinates in the volume D for which $\rho(y|t)$ has nonzero values. The jth component of the position vector of a particle is written as y^j, where $j = 1, 2, 3$. Indices 1, 2, 3 signify space components. The symbol t plays the role of time in nonrelativistic theory. In this formalism, nonrelativistic time t is an evolution parameter that conditions the probability density $\rho(y|t)$.

According to probability theory, $\rho(y|t)$ must be positive-definite and normalizable; thus

$$\rho(y|t) \geq 0$$

and

$$\int_D \rho(y|t)\,dy = 1, \, dy = d^3y = dy^1 dy^2 dy^3$$

Conservation of probability implies the continuity equation

$$\frac{\partial \rho}{\partial t} + \sum_{j=1}^{3} \frac{\partial \rho V^j}{\partial y^j} = 0$$

where the term ρV^j represents the jth component of probability flux, and V^j is a velocity vector.

For ρ to be differentiable and nonnegative, its derivative must satisfy $\partial \rho / \partial t = 0$ if $\rho = 0$ [Collins, 1977, 1979]. The positive-definite requirement for $\rho(y|t)$ is satisfied by writing $\rho(y|t)$ in the Born representation; thus

$$\rho(y|t) = \psi^*(y, t)\psi(y, t)$$

The scalar eigenfunctions ψ can be written as

$$\psi(y, t) = [\rho(y|t)]^{1/2} \exp[i\xi(y, t)]$$

where $\xi(y, t)$ is a real scalar function.

The Schrödinger equation is derived by expressing the velocity $\{V^j\}$ as

$$V^j(y, t) = \frac{1}{m}\left[\hbar \frac{\partial \xi(y, t)}{\partial y^j} - \frac{e}{c}A^j(y, t)\right]$$

We assume the vector A^j and consequently V^j are real functions. Using these assumptions, we can derive the field equation

$$i\hbar \frac{\partial \Psi}{\partial t} = \frac{1}{2m}\sum_{a=1}^{3} \pi^j \pi_j \Psi + U\Psi$$

where U is Hermitian. The operators π^j and p^j are defined by

$$\pi^j \equiv p^j - \frac{e}{c}A^j, \, p^j \equiv \frac{\hbar}{i}\frac{\partial}{\partial y^j}$$

Given the form of the equations, the vector A^j is identified as the electromagnetic vector potential acting on the particle. Additional interactions may be incorporated via the potential function U. For more information, the reader should see Fröhner's [1998] relatively recent discussion of the relationship between probability theory and quantum mechanics. A similar derivation can be used to obtain equations that behave like relativistic wave equations [Fanchi, 1993]. An example of one such relativistic equation is discussed in Section 12.3.

17.4   EXAMPLE PROBABILITY DISTRIBUTIONS

Many probability distributions have been developed to describe a variety of systems ranging from social to scientific. A few probability distributions are outlined below to illustrate their form and function. For more details, the reader should consult the references, such as Walpole and Myers [1985], Larsen and Marx [1985], Kurtz [1991], Blaisdell [1998], Miller and Miller [1999], and Kreyszig [1999].

Discrete Case

Uniform Distribution   The discrete random variable X has a uniform distribution if its probability function is

$$P(X = x) = f(x) = \frac{1}{n}, \, x = x_1, x_2, \ldots, x_n$$

The mean and variance of the uniform distribution are

$$\text{Mean} = \mu = \frac{n+1}{2}$$

$$\text{Variance} = \sigma^2 = \frac{n^2 - 1}{12}$$

Example: The probability distribution for the roll of a die with 6 sides is a uniform distribution with $n = 6$, $\mu = 3.5$, and $\sigma = 1.7$.

Binomial Distribution   The discrete random variable X has a binomial distribution if its probability function is

$$P(X = x) = f(x) = \binom{n}{x}\theta^x(1 - \theta)^{n-x}, \, x = 0, 1, 2, \ldots, n$$

with the binomial coefficient

$$\binom{n}{x} = \frac{n!}{x!(n-x)!}$$

The mean and variance of the binomial distribution are

$$\text{Mean} = \mu = n\theta$$

$$\text{Variance} = \sigma^2 = n\theta(1-\theta)$$

The binomial distribution arises when the outcomes of a trial can be classified as one of two possible events. In this case n is the number of independent trials and X is the number of times that a given event occurs. The binomial distribution gives the probability that x successes will occur out of n independent trials given that θ is the probability of success for one trial.

Geometric Distribution The discrete random variable X has a geometric distribution if its probability function is

$$P(X = x) = f(x) = \theta(1-\theta)^{x-1}, \, x = 1, 2, 3, \dots,$$

The mean and variance of the geometric distribution are

$$\text{Mean} = \mu = \frac{1}{\theta}$$

$$\text{Variance} = \sigma^2 = \frac{1-\theta}{\theta^2}$$

The geometric distribution can be used to determine the probability that a given event will occur on a particular trial, assuming that a trial is performed repeatedly and that each trial is independent. In this case, θ is the probability that a given event will occur on a single trial, and $X = x$ is the number of trials required to obtain the given event.

Example: The probability of rolling a 1 using a single die with six sides is $1/6$. If you are rolling the die until you get a 1, the probability that you will have to roll five times is found from the geometric distribution to be

$$P(5) = \frac{1}{6}\left(1 - \frac{1}{6}\right)^{5-1} = 0.08$$

Poisson Distribution   The discrete random variable X has a Poisson distribution if its probability function is

$$P(X = x) = f(x) = \frac{e^{-\lambda}\lambda^x}{x!}, \lambda > 0, x = 0, 1, \ldots,$$

The mean and variance of the Poisson distribution are

$$\text{Mean} = \mu = \lambda$$

$$\text{Variance} = \sigma^2 = \lambda$$

The Poisson distribution is the limiting case of the binomial distribution when $n \to \infty$ and $x \to 0$ such that the product $nx \to$ constant λ.

Hypergeometric Distribution   The discrete random variable X has a hypergeometric distribution if its probability function is

$$P(X = x) = f(x) = \frac{\binom{k}{x}\binom{N-k}{n-x}}{\binom{N}{n}}, x = 0, 1, 2, \ldots, [n, k]$$

where $[n,k]$ means the smaller of the two numbers n, k. The mean and variance of the hypergeometric distribution are

$$\text{Mean} = \mu = \frac{kn}{N}$$

$$\text{Variance} = \sigma^2 = \frac{k(N-k)\, n\, (N-n)}{N^2(N-1)}$$

The hypergeometric distribution arises when elements are drawn from a set without replacement. The random variable X is the number of times that an element of a given type is drawn.

Example: Suppose a box contains 6 blue balls and 8 red balls. Ten balls will be taken from the box without replacement. If we want to know the probability that 5 of the balls will be blue, we use the hypergeometric function. In this case N is the number of balls in the original set ($N = 14$); n is the number of balls drawn ($n = 10$); k is the number of blue balls in the original set ($k = 6$); and X is the number of balls that will be drawn ($X = x = 5$). The probability of obtaining

5 blue balls when drawing 10 balls from the box is

$$P(5) = \frac{\binom{6}{5}\binom{14-6}{10-5}}{\binom{14}{10}} = \frac{(6)(42)}{(1001)} = 0.252$$

Continuous Case

Uniform Distribution   The continuous random variable X has a uniform distribution if its probability density is

$$p(x) = \begin{cases} \dfrac{1}{\beta - \alpha} & \text{for } \alpha < x < \beta \\ 0 & \text{otherwise} \end{cases}$$

where α and β are parameters with $\alpha < \beta$. The mean and variance of the uniform distribution are

$$\text{Mean} = \mu = \frac{\alpha + \beta}{2}$$

$$\text{Variance} = \sigma^2 = \frac{(\beta - \alpha)^2}{12}$$

Normal Distribution   The continuous random variable X has a normal or Gaussian distribution if its probability density is

$$p(x) = \frac{1}{\sqrt{2\pi}\sigma} e^{-(x-\mu)^2/2\sigma^2}, \quad -\infty < x < \infty$$

where μ and σ are parameters. For the normal distribution, the parameters μ and σ are also the mean and standard deviation of the random variable X, respectively:

$$\text{Mean} = \mu$$

$$\text{Variance} = \sigma^2$$

The normal distribution is useful for describing very large sample groups. According to the *central limit theorem*, many probability distributions approach the normal distribution as the number of elements in the sample group approaches infinity.

The cumulative probability for the normal distribution is

$$F(x) = \int_{-\infty}^{x} \frac{1}{\sqrt{2\pi}\sigma} e^{-(x-\mu)^2/2\sigma^2} dx$$

The standard normal distribution has a mean of zero and variance of one. If we set $z = (x - \mu)/\sigma$ in $F(x)$, we obtain the cumulative standard normal distribution.

EXERCISE 17.6: Determine the expectation value $E(x)$ using the probability density for the normal distribution.

Gamma Distribution   The continuous random variable X has a gamma distribution if its probability density is

$$p(x) = \frac{1}{\Gamma(\alpha + 1)\,\beta^{\alpha+1}}\, x^{\alpha} e^{-x/\beta},\ 0 < x < \infty$$

where α and β are parameters with $\alpha > -1$ and $\beta > 0$. The gamma function $\Gamma(z)$ is defined as the integral

$$\Gamma(z) = \int_{0}^{\infty} y^{z-1} e^{-y} dy$$

for $z > 0$. The mean and variance of the gamma distribution are

$$\text{Mean} = \mu = \beta(\alpha + 1)$$
$$\text{Variance} = \sigma^2 = \beta^2(\alpha + 1)$$

If $\alpha = 0$, the gamma distribution simplifies to the exponential distribution.

Exponential Distribution   The continuous random variable X has an exponential distribution if its probability density is

$$p(x) = \frac{1}{\theta} e^{-x/\theta},\ x > 0$$

where θ is a parameter and $\theta > 0$. The mean and variance of the exponential distribution are

$$\text{Mean} = \mu = \theta$$
$$\text{Variance} = \sigma^2 = \theta^2$$

The exponential distribution is often used to describe the amount of a radioactive substance remaining at a given time.

CHAPTER 18

STATISTICS

Fundamental concepts used in descriptive statistics are reviewed here, beginning with the presentation of the relationship between probability and frequency. This is followed by a discussion of the statistics of grouped and ungrouped data. Several statistical coefficients for describing a distribution of data are defined. Commonly used formulas for curve fitting and regression are then summarized.

18.1 PROBABILITY AND FREQUENCY

The relationship between probability and frequency is based on the *law of large numbers* of probability theory. Consider an event $E = E_j$ that is an element of a set of J events. Suppose that out of N trials the event E_j is observed to occur N_j times. The observed frequency f_j of the event is

$$f_j = \frac{N_j}{N}$$

The law of large numbers states that the probability P_j that the event E_j will occur is related to the frequency of occurrence f_j by the limit

$$\lim_{N \to \infty} |P_j - f_j| = 0$$

Math Refresher for Scientists and Engineers, Third Edition By John R. Fanchi
Copyright © 2006 John Wiley & Sons, Inc.

In theory, the number of trials should approach infinity for the law of large numbers to apply. In practice, the law of large numbers means that the frequency f_j can be used for the probability P_j that the event E_j will occur when the number of trials is sufficiently large. The term "sufficiently large" depends on the situation of interest.

Example: Suppose a six-sided die with numbers $\{1, 2, 3, 4, 5, 6\}$ on the sides is rolled 100 times and the number 1 appears on the top of the die 18 times. The observation of the number 1 on the top of the rolled die is event E_1 of the set of events $\{E_1, \ldots, E_6\} = \{1, \ldots, 6\}$ with $J = 6$ events. The frequency of obtaining the number 1 is $f_1 = N_1/N = 18/100$ and the corresponding probability is $P_1 \approx f_1 = 0.18$. As the number of die rolls increases, the probability will approach $1/6$ for an unbiased die.

Data Grouping

Suppose we have a set with n data elements $\{x_i : i = 1, 2, \ldots, n\}$. Each data element x_i has a value, and the range of values is the difference between the largest and smallest values in the set of elements. The set of elements is considered ungrouped if each element in the set is treated as a separate element rather than as a member of a subset with two or more elements. The elements in the set can be grouped by collecting elements in classes.

A class is a collection of elements having a value in a subinterval of the range of values. The number of classes is subjective, but is often somewhere between 5 classes and 20 classes [Kurtz, 1991]. Each class has the same interval or uniform width c. The lower value of the class is the lower class limit L. If the class does not have a lower or an upper limit, it is an open class. The number of data points in the class is the class frequency f, as illustrated in Table 18.1. Cumulative frequency in a class interval is the sum of the frequencies of all class intervals up to and including the current class interval. The cumulative frequency for the last class is the total number of elements in the set.

TABLE 18.1   Example of Grouped Statistical Data

Class Interval	Number of Occurrences	Frequency	Cumulative Frequency
0.05 to 0.10	2	0.0317	0.0317
0.10 to 0.15	4	0.0635	0.0952
0.15 to 0.20	15	0.2381	0.3333
0.20 to 0.25	24	0.3810	0.7143
0.25 to 0.30	14	0.2223	0.9366
0.30 to 0.35	2	0.0317	0.9683
0.35 to 0.40	2	0.0317	1.0000
Total	63	1.0000	

The actual value of each element disappears in a class and is replaced by a value that is uniformly distributed. Thus the rth value in the class X_r is

$$X_r = L + \frac{r}{f}c$$

The midpoint or mark of a class is an element with the value $L + (c/2)$; that is, the midpoint of the class is halfway between the lower and upper limits of the class. If an element has the value L corresponding to the upper limit of one class and the lower limit L of an adjacent class, it is necessary to specify which class will be assigned elements with values equal to L.

Example: Suppose a class of measurements has an interval ranging from 0.20 to 0.25. The interval "0.20 to 0.25" in this case is actually the interval "0.20 up to but not including 0.25." The number of observations in the class is 10. The fourth value in the class is $X_4 = 0.20 + (4/10)\ 0.05 = 0.22$. The mark or midpoint of the class is 0.225. If a measurement has the value 0.25, it will be assigned to the class of measurements with an interval ranging from 0.25 to 0.30.

Frequency Diagrams

It is often useful to provide a visual representation of the frequency of occurrence of data. This can be done by plotting the value of a datum along the horizontal axis and its frequency of occurrence along the vertical axis. If the frequency of occurrence is divided by the total number of data points, the frequency diagram becomes a diagram showing a probability distribution, subject to the conditions associated with the law of large numbers described earlier in this section.

Grouped data are represented by histograms. The histogram is a presentation of statistical data as rectangles. The width of the rectangle is the width of the class and the height of the rectangle is the frequency of the class. For further discussion of techniques for presenting statistical data, see such references as Kurtz [1991] and Kreyszig [1999].

EXERCISE 18.1: Prepare a histogram of the grouped data in Table 18.1.

18.2 UNGROUPED DATA

Several different *means* may be defined for a set of data elements. The arithmetic mean for a set of ungrouped data with n elements is

$$\bar{x} = \frac{1}{n}\sum_{i=1}^{n} x_i = \frac{x_1 + x_2 + \cdots + x_n}{n}$$

where x_i denotes the values of the ith element. A weighted arithmetic mean can be defined by associating a weighting factor $w_i \geq 0$ with each value. The weighted arithmetic mean is then

$$\bar{x}_w = \frac{\sum_{i=1}^{n} w_i x_i}{\sum_{i=1}^{n} w_i} = \frac{w_1 x_1 + w_2 x_2 + \cdots + w_n x_n}{w_1 + w_2 + \cdots + w_n}$$

The geometric mean for ungrouped data is

$$\bar{x}_G = \sqrt[n]{x_1 \cdot x_2 \cdots x_n}$$

The harmonic mean for ungrouped data is

$$\bar{x}_H = \frac{n}{\sum_{i=1}^{n} \frac{1}{x_i}} = \frac{n}{\dfrac{1}{x_1} + \dfrac{1}{x_2} + \cdots + \dfrac{1}{x_n}}$$

The relationship between arithmetic, geometric, and harmonic means is

$$\bar{x}_H \leq \bar{x}_G \leq \bar{x}$$

where the equality signs apply only if all sample values are identical.

A mode M_o of a sample of size n is the value that occurs with greatest frequency; that is, it is the most common value. The class with the highest frequency is the modal class. A mode may not exist, and if it does exist it may not be unique. For example, if a distribution has two values that occur with the same maximum frequency, the distribution is said to be *bimodal*.

The median is often used as a representative value of a set of data. If a sample with n elements is arranged in ascending order of magnitude, then the median M_d is given by the $(n + 1)/2$ value. When n is odd, the median is the middle value of the set of ordered data; when n is even, the median is the arithmetic mean of the two middle values of the set of ordered data.

The mean absolute deviation for ungrouped data is

$$\text{M.A.D.} = \frac{1}{n} \sum_{i=1}^{n} |x_i - \bar{x}|$$

where $\bar{x}$ is the arithmetic mean. The standard deviation for ungrouped data is the root mean square of the deviations from the arithmetic mean; thus

$$s = \sqrt{\frac{\sum_{i=1}^{n} (x_i - \bar{x})^2}{n}}$$

The variance is the square of the standard deviation. The root mean square for ungrouped data is

$$\text{R.M.S.} = \left[\frac{1}{n} \sum_{i=1}^{n} x_i^2 \right]^{1/2}$$

18.3 GROUPED DATA

The total number of observations in a frequency distribution having k classes with corresponding class frequencies $\{f_i : i = 1, 2, \ldots, k\}$ is

$$n = \sum_{i=1}^{k} f_i$$

The arithmetic mean for grouped data is

$$\bar{x} = \frac{1}{n} \sum_{i=1}^{k} f_i x_i = \frac{f_1 x_1 + f_2 x_2 + \cdots + f_k x_k}{n}$$

where $\{x_i : i = 1, 2, \ldots, k\}$ is the set of class marks. The geometric mean for grouped data is

$$\bar{x}_G = \sqrt[n]{x_1^{f_1} \cdot x_2^{f_2} \cdots x_k^{f_k}}$$

The harmonic mean for grouped data is

$$\bar{x}_H = \frac{n}{\sum_{i=1}^{k} \dfrac{f_i}{x_i}} = \frac{n}{\dfrac{f_1}{x_1} + \dfrac{f_2}{x_1} + \cdots + \dfrac{f_k}{x_k}}$$

The relationship between arithmetic, geometric, and harmonic means is

$$\bar{x}_H \leq \bar{x}_G \leq \bar{x}$$

where the equality signs apply only if all sample values are identical.

EXERCISE 18.2: What is the mean of the grouped data in Table 18.1?

The modal class is the class with the highest frequency. The mode M_o for grouped data is

$$M_o = L + C \frac{d_1}{d_1 + d_2}$$

where L is the lower limit of the modal class, C is the width of the modal class, d_1 is the difference between the frequency of the modal class and that of the preceding class, and d_2 is the difference between the frequency of the modal class and that of the succeeding class.

Example: The mode for the grouped data in Table 18.1 is $0.20 + 0.05(24 - 15)/[(24 - 15) + (24 - 14)] = 0.224$. The modal class has the interval 0.20 to 0.25.

The median of grouped data is element $n/2$ of a set of n values. If n is even, the median is a real element, otherwise it is hypothetical. The median class is the class containing the median. The median for grouped data is

$$M_d = L + c\frac{(n/2) - F_c}{f_m}$$

where n is the number of classes, L is the lower limit of the median class, c is the width of the median class, F_c is the sum of the frequencies of all classes lower than the median class, and f_m is the frequency of the median class.

The median is always between the mode and the mean. If the mode is to the left of the mean on a frequency diagram, the distribution is skewed to the left. If the mode is to the right of the mean on a frequency diagram, the distribution is skewed to the right.

Example: The median for the grouped data in Table 18.1 is between elements 31 and 32. Using the above equation for the median gives $0.20 + 0.05(31.5 - 21)/24 = 0.222$. The median class has the interval 0.20 to 0.25. The mean of the grouped data in Table 18.1 is calculated in Exercise 18.2 to be 0.221. Comparing the mode and the mean, we find the mode (0.237) is greater than the mean (0.221), which places the mode to the right of the mean on a frequency diagram. Therefore the distribution of grouped data in Table 18.1 is skewed to the right.

The mean absolute deviation for grouped data is

$$\text{M.A.D.} = \frac{1}{n}\sum_{i=1}^{k}f_i|x_i - \bar{x}|$$

where $\bar{x}$ is the arithmetic mean. The standard deviation for grouped data is the root mean square of the deviations from the arithmetic mean; thus

$$s = \sqrt{\frac{\sum_{i=1}^{k}f_i(x_i - \bar{x})^2}{n}}$$

The variance is the square of the standard deviation. The root mean square for grouped data is

$$\text{R.M.S.} = \left[\frac{1}{n}\sum_{i=1}^{k}f_i x_i^2\right]^{1/2}$$

18.4 STATISTICAL COEFFICIENTS

A number of coefficients are used to describe a statistical distribution. Some of the more common coefficients are defined here. The simplest coefficients are defined in terms of means, standard deviations, and variances. More complex coefficients are expressed in terms of moments of the distribution.

The *coefficient of variation* is defined as

$$V = \frac{100s}{\bar{x}}$$

where $\bar{x}$ is the mean and s the standard deviation of the sample. The *standardized variable* or *standard score* associated with observation x_i is defined as

$$z = \frac{x_i - \bar{x}}{s}$$

where $\bar{x}$ is the mean and s the standard deviation of the sample.

The concept of moments was introduced previously to describe probability distributions. A similar concept applies to statistical distributions. It is presented here in terms of ungrouped and grouped data.

For ungrouped data, the rth moment about the origin is

$$m'_r = \frac{1}{n} \sum_{i=1}^{n} x_i^r$$

The rth moment about the mean is

$$m'_r = \frac{1}{n} \sum_{i=1}^{n} (x_i - \bar{x})^r$$

For grouped data, the rth moment about the origin is

$$m_r = \frac{1}{n} \sum_{i=1}^{k} f_i x_i^r$$

The rth moment about the mean $\bar{x}$ is

$$m_r = \frac{1}{n} \sum_{i=1}^{k} f_i (x_i - \bar{x})^r$$

Example: The second moment is the variance of the statistical distribution for both grouped and ungrouped data. Consequently, the standard deviation of a statistical distribution is the square root of the second moment.

Given the concept of moments, we can calculate several coefficients describing the shape of the statistical distribution. The coefficients of skewness is a measure of the symmetry of the distribution. If the distribution is unsymmetric about the mode, it is said to be *skewed* and the degree of asymmetry is referred to as *skewness*. The coefficient of skewness is

$$\alpha_3 = \frac{m_3}{(m_2)^{3/2}}$$

where m_2 and m_3 are the second and third moments about the mean of the sample, respectively. If the magnitude of α_3 is close to zero, the distribution displays a high degree of symmetry about the mode. If the magnitude of α_3 is greater than one, the distribution is highly skewed.

Another measure of the shape of a statistical distribution is the *coefficient of kurtosis*. The coefficient of kurtosis is

$$\alpha_4 = \frac{m_4}{(m_2)^2}$$

where m_2 and m_4 are the second and fourth moments about the mean of the sample, respectively. It provides an indication of the width of the statistical distribution relative to its height. If we note that the standard deviation of the statistical distribution is the square root of the second moment, we see that a relatively small standard deviation implies a relatively large coefficient of kurtosis if the fourth moment is constant. Thus a large coefficient of kurtosis implies a relatively narrow statistical distribution. For comparison purposes, the coefficient of kurtosis for the normal probability distribution is 3 [Kurtz, 1991, p. 830].

Example: A statistical distribution is symmetric about the mode (coefficient of skewness = 0) and has a coefficient of kurtosis of 2. It therefore has a narrower distribution than the normal probability distribution. It must also have a higher peak at the mode to ensure that the area under the curve is normalized.

18.5 CURVE FITTING, REGRESSION, AND CORRELATION

Regression is the estimation of one variable y from another variable x. The relationships presented in this section apply to a set of n ordered pairs $\{(x_i, y_i): i = 1, 2, \ldots, n\}$ of independent variable x and dependent variable y. The ordered pair (x_i, y_i) is assumed to be a random sample from a bivariate normal distribution. The variable x is the predictor or regressor variable, and the variable y is the predicted or regressed variable. Regression is achieved by minimizing the deviation of a curve fit to a set of data as outlined below.

Curve Fitting

It is often useful to fit a smooth curve through data. Several smooth curves may be used. The simplest is the straight line $y = mx + b$, where m is the slope and b is the y intercept. For a straight line fit by the method of least squares (Chapter 3), the values b and m are obtained by solving the algebraic equations

$$nb + m \sum_i x_i = \sum_i y_i$$

and

$$b \sum_i x_i + m \sum_i x_i^2 = \sum_i x_i y_i$$

The solutions of these equations are given in Chapter 3. Fitting a straight line to a set of data may appear to have limited utility; however, it is possible to recast a number of useful nonlinear functions in a form that is suitable for analysis by least squares.

Example: The exponential curve $y = ab^x$ can be recast in linear form by taking the logarithm

$$\log y = \log a + (\log b)x$$

For an exponential curve fit by the method of least squares, the values $\log a$ and $\log b$ are obtained by fitting a straight line to the set of ordered pairs $\{(x_i, \log y_i)\}$.

Similar regression procedures can be applied to polynomial functions of the form

$$y = b_0 + b_1 x + b_2 x^2 + \cdots + b_m x^m$$

For a polynomial function fit by the method of least squares, the value of the coefficients $b_0, b_1, \ldots, b_m$ are obtained by solving a system of $m + 1$ algebraic equations. The reader should consult the references for more details.

EXERCISE 18.3: Show that the power function $y = ax^b$ can be recast in linear form.

Regression and Correlation

Least squares analysis is an example of a more general technique known as regression. Regression is used to determine causal relationships between observed variables. For example, simple linear regression seeks to find a relationship for variable y given variable x; thus

$$E(y|x) = b + mx$$

where $E(y|x)$ is the mean of the distribution of y for a given x. The term *simple* implies that y depends on only one variable. If y depended on more than one variable, the regression would be called a *multiple* regression. The form of the relationship between variables x and y is often referred to as a *regression model*. In this case, we have a linear regression model. The regression coefficients b and m are determined by minimizing the deviations of the observed y values from the straight line.

A measure of the validity of the regression model is the standard error of estimate. For simple linear regression, the standard error of estimate (SEE) is

$$s_e = \sqrt{\frac{\sum_i [y_i - (b + mx_i)]^2}{n}}$$

A large value of SEE implies large deviations from the assumed linear relationship between ordered pairs. By contrast, a small value of SEE implies that the set of ordered pairs closely follows a linear relationship.

The total variation of the dependent variable y from the mean is given by the sum of squares total:

$$SS_T = \sum_i (y_i - \bar{y})^2$$

For comparison, the variation of the regressed value of the dependent variable y from the mean is given by the sum of squares regression:

$$SS_R = \sum_i [(b + mx_i) - \bar{y}]^2$$

The ratio of the sum of squares regression to the sum of squares total is the goodness-of-fit:

$$R^2 = \frac{SS_R}{SS_T}$$

The square root of the goodness-of-fit variable R^2 gives the correlation coefficient r; thus

$$r = \sqrt{R^2}$$

An alternate but equivalent form of r is the product-moment formula [Kurtz, 1991, p. 10.39; Miller and Miller, 1999, p. 453]

$$r = \frac{\sum_i x_i y_i}{\sqrt{\left(\sum_i x_i^2\right)\left(\sum_i y_i^2\right)}} = \frac{\sigma_{xy}}{\sigma_x \sigma_y}$$

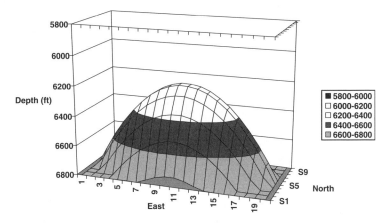

Figure 18.1 Depth to top of anticlinal structure.

The terms σ_x and σ_y are the standard deviations for the x and y distributions, and σ_{xy} is the covariance of x and y.

The correlation coefficient r is a dimensionless measure of the validity of a linear relationship between two variables. It can range from $+1$ to -1. The sign of r is the same as the sign of the slope m of the linear regression model. The value $|r| = 1$ says that all points of the ordered pair lie on a straight line. If $r = 0$, there is no linear relationship. A value of $r = 0$ is possible even if two variables are strongly correlated if the correlation is nonlinear. On the other hand, a value of $|r| \approx 1$ does not imply causality between two variables, but only statistical correlation [Mendenhall and Beaver, 1991, p. 424].

Application: Trend Surface Analysis. A technique for determining the spatial distribution of a property by computer is to fit a surface through the control points. Control points are values of the property at particular locations in space. This technique is referred to here as *trend surface analysis*. Linear trend surface analysis uses regression to fit a line through all control point values in a given direction. The regression model for linear trend surface analysis is

$$V_{\text{obs}} = a_0 + a_1 x_{\text{loc}} + a_2 y_{\text{loc}}$$

where V_{obs} is the observed value at the control point, and $\{x_{\text{loc}}, y_{\text{loc}}\}$ are the $\{x$ axis, y axis$\}$ locations of the control point. At least six control points are needed to determine the regression coefficients $\{a_0, a_1, a_2, a_3, a_4, a_5\}$. Quadratic trend surface analysis requires more control points than linear trend surface analysis to fit the larger set of regression coefficients.

Quadratic trend surface analysis can fit a curved surface to data and is therefore useful in representing geologic structures such as anticlines or synclines. An example of an anticlinal surface obtained from trend surface analysis is illustrated in Figure 18.1.

CHAPTER 19

SOLUTIONS TO EXERCISES

S1 ALGEBRA

EXERCISE 1.1: Given $ax^2 + bx + c = 0$, find x in terms of a, b, c.

Solution: We use the method of completing the square to derive the quadratic formula. Write

$$ax^2 + bx = -c$$

Divide by a to get

$$x^2 + \frac{b}{a}x = \frac{-c}{a}$$

The left-hand side is made a perfect square by adding $(b/2a)^2$ to both sides; thus

$$x^2 + \frac{b}{a}x + \left(\frac{b}{2a}\right)^2 = \left(x + \frac{b}{2a}\right)^2 = -\frac{c}{a} + \frac{b^2}{4a^2}$$

Rearranging the right-hand side gives

$$\left(x + \frac{b}{2a}\right)^2 = \frac{b^2 - 4ac}{4a^2}$$

Math Refresher for Scientists and Engineers, Third Edition By John R. Fanchi
Copyright © 2006 John Wiley & Sons, Inc.

Taking the square root of both sides and rearranging yields

$$x = -\frac{b}{2a} \pm \frac{\sqrt{b^2 - 4ac}}{2a}$$

EXERCISE 1.2: (a) Expand and simplify $(x + y)^2$, $(x - y)^2$, $(x + y)^3$, $(x + y + z)^2$, $(ax + b) \cdot (cx + d)$, and $(x + y)(x - y)$. (b) Factor $3x^3 + 6x^2y + 3xy^2$.

Solution (a):

$$(x + y)^2 = x^2 + 2xy + y^2$$
$$(x - y)^2 = x^2 - 2xy + y^2$$
$$(x + y)^3 = (x + y)^2(x + y)$$
$$= x^3 + 3x^2y + 3xy^2 + y^3$$
$$(x + y + z)^2 = x^2 + y^2 + z^2 + 2xy + 2xz + 2yz$$
$$(ax + b)(cx + d) = acx^2 + (bc + ad)x + bd$$
$$(x + y)(x - y) = x^2 - y^2$$

Solution (b):

$$3x^3 + 6x^2y + 3xy^2 = 3x(x^2 + 2xy + y^2) = 3x(x + y)^2$$

EXERCISE 1.3: Let $y = \log_a x$. Change the base of the logarithm from base a to base b, then set $a = 10$ and $b = 2.71828 \approx e$ and simplify.

Solution: We note that

$$\log_a x = y \tag{i}$$

and

$$x = a^y$$

To change the base from a to b, we scale the base a by writing

$$x = \left(\frac{ab}{b}\right)^y$$

Taking the logarithm to the base b of x gives

$$\log_b x = y[\log_b a + \log_b b - \log_b b] = y\log_b a$$

Solving for y yields

$$y = \frac{\log_b x}{\log_b a} = \log_a x \tag{ii}$$

where we have used Eq. (i). From Eq. (ii) we obtain the equality

$$\log_b x = (\log_a x)(\log_b a)$$

which shows how to change the logarithm from base a to base b.

Suppose $a = 10$ and $b = 2.71828 \approx e$. Then

$$\log_e x \equiv \ln x = \log_{10} x \log_e 10$$
$$= \log_{10} x \ln 10$$
$$= 2.3026 \log_{10} x$$

EXERCISE 1.4: What are the odds of winning a lotto? To win a lotto, you must select 6 numbers without replacement from a set of 42. Order is not important. Note that the combination of n different things taken k at a time without replacement is

$$\binom{n}{k} = \frac{n!}{k!(n-k)!}$$

Solution: For this lotto, $n = 42$ and $k = 6$; thus

$$\binom{42}{6} = \frac{42!}{6!(36)!} = \frac{42 \cdot 41 \cdot 40 \cdot 39 \cdot 38 \cdot 37 \cdot 36!}{6! \, 36!} = \frac{42 \cdot 41 \cdot 40 \cdot 39 \cdot 38 \cdot 37}{6!}$$

$$= \frac{3.78 \times 10^9}{720} = 5.25 \times 10^6$$

Notice that $42!/36! = 3.78 \times 10^9$. The odds of winning this lotto are approximately 1 in 5.25×10^6.

EXERCISE 1.5: Calculate $42!/36!$ using Stirling's approximation for both $42!$ and $36!$.

Solution: From Stirling's approximation

$$n! \approx e^{-n} n^n \sqrt{2\pi n}$$

we find

$$42! \approx e^{-42} 42^{42} \sqrt{2\pi(42)}$$
$$36! \approx e^{-36} 36^{36} \sqrt{2\pi(36)}$$

Dividing $42!$ by $36!$ gives

$$\frac{42!}{36!} \approx \frac{e^{-42} 42^{42} \sqrt{2\pi(42)}}{e^{-36} 36^{36} \sqrt{2\pi(36)}} \approx \frac{e^{-42}}{e^{-36}} \frac{42^{42}}{36^{36}} \sqrt{\frac{42}{36}}$$

This is further simplified by performing the arithmetic

$$\frac{42!}{36!} \approx e^{-6}42^6\left(\frac{42}{36}\right)^{36}(1.0801)$$

$$\approx e^{-6}42^6(1.1667)^{36}(1.0801)$$

$$\approx e^{-6}42^6(257.35)\,(1.0801)$$

$$\approx 3.78 \times 10^9$$

which agrees with the exact calculation in Exercise 1.4.

EXERCISE 1.6: Factor z_1/z_2 by calculating

$$\frac{z_1}{z_2}\frac{z_2^*}{z_2^*} = \frac{z_1 z_2^*}{z_2 z_2^*}$$

Solution:

$$\frac{z_1}{z_2} = \frac{x_1 + iy_1}{x_2 + iy_2} = \frac{x_1 + iy_1}{x_2 + iy_2}\frac{(x_2 - iy_2)}{(x_2 - iy_2)}$$

$$= \frac{x_1x_2 + y_1y_2}{x_2^2 + y_2^2} + i\left(\frac{y_1x_2 - x_1y_2}{x_2^2 + y_2^2}\right)$$

EXERCISE 1.7: Suppose $z = x + iy$ and $z^* = x - iy$. Find x, y in terms of z, z^*.

Solution: We find x and y by taking the sum and difference of z, z^*. Thus

$$z + z^* = x + iy + x - iy = 2x$$

which implies

$$x = \tfrac{1}{2}(z + z^*)$$

and

$$z - z^* = x + iy - (x - iy) = 2iy$$

which implies

$$y = \frac{1}{2i}(z - z^*)$$

EXERCISE 1.8: Given

$$g = \frac{i}{\varepsilon + i\lambda/2}$$

evaluate g^*g.

Solution

$$g^*g = \frac{-i}{\varepsilon - \dfrac{i\lambda}{2}} \; \frac{i}{\varepsilon + \dfrac{i\lambda}{2}} = \frac{-i^2}{\varepsilon^2 - \dfrac{i^2\lambda^2}{4}} = \frac{1}{\varepsilon^2 + \dfrac{\lambda^2}{4}}$$

EXERCISE 1.9: Write $1/(s^2 + \omega^2)$ in terms of partial fractions.

Solution: Use Eq. (1.8.6) to expand $1/(s^2 + \omega^2)$ as

$$\frac{1}{s^2 + \omega^2} = \frac{A_1}{a_1 s + b_1} + \frac{A_2}{a_2 s + b_2} \tag{i}$$

where the numerator is $r(s) = 1$. The denominator factors into the terms

$$D(s) = s^2 + \omega^2 = (s + i\omega)(s - i\omega) \tag{ii}$$

The coefficient values for Eq. (i) are

$$\begin{aligned} a_1 &= 1, b_1 = i\omega \\ a_2 &= 1, b_2 = -i\omega \end{aligned} \tag{iii}$$

Multiply Eq. (i) by the denominator $D(s)$:

$$\begin{aligned} 1 &= A_1 \frac{s^2 + \omega^2}{s + i\omega} + A_2 \frac{s^2 + \omega^2}{s - i\omega} \\ &= A_1(s - i\omega) + A_2(s + i\omega) \\ &= (A_1 + A_2)s + i\omega(A_2 - A_1) \end{aligned} \tag{iv}$$

The coefficients of s^n, $n = 0, 1$ in Eq. (iv) satisfy the equalities

$$\begin{aligned} i\omega(A_2 - A_1) &= 1 \\ A_1 + A_2 &= 0 \end{aligned} \tag{v}$$

Solving for A_1, A_2 gives

$$A_1 = -\frac{1}{2i\omega}, A_2 = \frac{1}{2i\omega} \tag{vi}$$

Substituting Eqs. (iii) and (vi) into (i) gives the partial fraction expansion

$$\frac{1}{s^2 + \omega^2} = -\frac{1}{2i\omega} \frac{1}{s + i\omega} + \frac{1}{2i\omega} \frac{1}{s - i\omega} \tag{vii}$$

EXERCISE 1.10: Suppose $r(x) = x + 3$, $D(x) = x^2 + x - 2$. Expand $r(x)/D(x)$ by the method of partial fractions.

Solution: The denominator $D(x)$ factors into the terms

$$D(x) = x^2 + x - 2 = (x+2)(x-1) \tag{i}$$

Use Eq. (1.8.6) to expand $r(x)/D(x)$ as

$$\frac{x+3}{x^2+x-2} = \frac{A_1}{a_1x+b_1} + \frac{A_2}{a_2x+b_2}$$
$$= \frac{A_1}{x+2} + \frac{A_2}{x-1} \tag{ii}$$

The coefficient values for Eq. (ii) are

$$a_1 = 1, b_1 = 2$$
$$a_2 = 1, b_2 = -1 \tag{iii}$$

Multiply Eq. (ii) by the denominator $D(x)$ to obtain

$$x + 3 = A_1(x-1) + A_2(x+2)$$
$$= (A_1 + A_2)x - A_1 + 2A_2 \tag{iv}$$

The coefficients of x^n, $n = 0, 1$ in Eq. (iv) satisfy the equalities

$$-A_1 + 2A_2 = 3$$
$$A_1 + A_2 = 1 \tag{v}$$

Solving for A_1, A_2 gives

$$A_1 = -\frac{1}{3}, A_2 = \frac{4}{3} \tag{vi}$$

Substituting Eqs. (iii) and (vi) into (ii) gives the partial fraction expansion

$$\frac{x+3}{x^2+x-2} = -\frac{1}{3}\frac{1}{x+2} + \frac{4}{3}\frac{1}{x-1} \tag{vii}$$

S2   GEOMETRY, TRIGONOMETRY, AND HYPERBOLIC FUNCTIONS

EXERCISE 2.1: Verify the following relations:

$$\sin^2 \alpha + \cos^2 \alpha = 1$$
$$1 + \tan^2 \alpha = \sec^2 \alpha$$

$$1 + \cot^2 \alpha = \csc^2 \alpha$$

$$\sin \alpha = \tan \alpha \cos \alpha$$

$$\cos \alpha = \cot \alpha \sin \alpha$$

$$\tan \alpha = \frac{\sin \alpha}{\cos \alpha}$$

by rearranging the trigonometric definitions and using the Pythagorean relation.

Solution: Recall the Pythagorean relation

$$a^2 + b^2 = c^2 \tag{i}$$

Many trigonometric identities can be proved by simple manipulation of Eq. (i). The identities in this exercise illustrate this procedure. Divide Eq. (i) by c^2 to get

$$\left(\frac{a}{c}\right)^2 + \left(\frac{b}{c}\right)^2 = \sin^2 \alpha + \cos^2 \alpha = 1$$

Divide Eq. (i) by a^2 to get

$$1 + \left(\frac{b}{a}\right)^2 = \left(\frac{c}{a}\right)^2$$

or

$$1 + \cot^2 \alpha = \csc^2 \alpha$$

Divide Eq. (i) by b^2 to get

$$\left(\frac{a}{b}\right)^2 + 1 = \left(\frac{c}{b}\right)^2$$

or

$$\tan^2 \alpha + 1 = \sec^2 \alpha$$

From the definitions of the trigonometric functions we obtain

$$\sin \alpha = \frac{a}{c} = \frac{a}{b}\frac{b}{c} = \tan \alpha \cos \alpha$$

$$\cos \alpha = \frac{b}{c} = \frac{b}{a}\frac{a}{c} = \cot \alpha \sin \alpha$$

$$\tan \alpha = \frac{a}{b} = \frac{a}{c}\frac{c}{b} = \frac{\sin \alpha}{\cos \alpha} = \sin \alpha \sec \alpha$$

EXERCISE 2.2: *RLC Circuit.* The alternating current $I(t)$ for a resistor–inductor–capacitor circuit with an applied electromotive force is

$$I(t) = -\frac{E_0\, S}{S^2 + R^2}\, \cos(\omega t) + \frac{E_0\, R}{S^2 + R^2}\, \sin(\omega t)$$

Use trigonometric relations to write $I(t)$ in the form

$$I(t) = I_0 \sin(\omega t - \delta)$$

where the maximum current I_0 and phase δ are expressed in terms of E_0, S, and R. A derivation of $I(t)$ is given in Chapter 11, Exercise 11.1.

Solution: Use the trigonometric identity $\sin(A - B) = \sin A \cos B - \cos A \sin B$ to expand

$$I_0 \sin(\omega t - \delta) = I_0(\sin \omega t \cos \delta - \cos \omega t \sin \delta)$$
$$= -I_0 \sin \delta \cos \omega t + I_0 \cos \delta \sin \omega t$$

Equate coefficients of $\cos \omega t$, $\sin \omega t$ to obtain the two equations

$$-I_0 \sin \delta = -\frac{E_0\, S}{S^2 + R^2}$$

$$I_0 \cos \delta = \frac{E_0\, R}{S^2 + R^2}$$

Take the ratio of equations to find δ:

$$\frac{\sin \delta}{\cos \delta} = \tan \delta = \frac{E_0\, S}{E_0\, R}$$

or

$$\tan \delta = \frac{S}{R}$$

Square both equations and sum to find I_0:

$$I_0^2 \cos^2 \delta + I_0^2 \sin^2 \delta = \frac{E_0^2\, R^2}{(S^2 + R^2)^2} + \frac{E_0^2\, S^2}{(S^2 + R^2)^2}$$

Combining terms and simplifying gives

$$I_0^2(\cos^2 \delta + \sin^2 \delta) = \frac{E_0^2(R^2 + S^2)}{(S^2 + R^2)^2}$$

or

$$I_0^2 = \frac{E_0^2}{(S^2 + R^2)}$$

Rearranging gives

$$E_0 = I_0\sqrt{S^2 + R^2}$$

Comparing this equation with Ohm's law ($V = IR$) shows that impedance $\sqrt{S^2 + R^2}$ acts like resistance for an alternating current (AC) system.

EXERCISE 2.3: Derive expressions for $\cos\alpha$, $\sin\alpha$, and $\tan\alpha$ in terms of exponential functions.

Solution: Recall from Euler's equation that

$$e^{i\alpha} = \cos\alpha + i\sin\alpha$$

and

$$e^{-i\alpha} = \cos(-\alpha) + i\sin(-\alpha) = \cos\alpha - i\sin\alpha$$

From these relations we readily obtain

$$\cos\alpha = \frac{e^{i\alpha} + e^{-i\alpha}}{2}$$

$$\sin\alpha = \frac{e^{i\alpha} - e^{-i\alpha}}{2i}$$

and

$$\tan\alpha = \frac{\sin\alpha}{\cos\alpha} = -i\left[\frac{e^{i\alpha} - e^{-i\alpha}}{e^{i\alpha} + e^{-i\alpha}}\right]$$

EXERCISE 2.4: Use the exponential forms of $\sinh u$ and $\cosh u$ to show that $\sinh(-u) = -\sinh u$ and $\cosh(-u) = \cosh u$.

Solution:
$$\sinh(-u) = \tfrac{1}{2}(e^{-u} - e^u) = -\tfrac{1}{2}(e^u - e^{-u}) = -\sinh u$$
$$\cosh(-u) = \tfrac{1}{2}(e^{-u} + e^u) = \tfrac{1}{2}(e^u + e^{-u}) = \cosh u$$

EXERCISE 2.5: Evaluate $\cosh^2 u - \sinh^2 u$.

Solution: Convert the relation to exponential form and then evaluate.

$$\cosh^2 u - \sinh^2 u = \left[\tfrac{1}{2}(e^u + e^{-u})\right]^2 - \left[\tfrac{1}{2}(e^u - e^{-u})\right]^2$$
$$= \tfrac{1}{4}\{e^{2u} + 2 + e^{-2u} - (e^{2u} - 2 + e^{-2u})\}$$
$$= \tfrac{1}{4}\{4\} = 1$$

EXERCISE 2.6: Let $z = x - vt$, $v = 1.0$ foot/day, $b = 0.1$ foot/day, and $D = 0.1$ foot^2/day. Plot $C(z)$ in the interval $0 \leq x \leq 40$ feet at times of 5.00 days, 6.11 days, and 7.22 days. Note that $C(z) = \frac{1}{2}[1 - \tanh(bz/4D)]$ and $0.0 \leq C(z) \leq 1.0$.

Solution

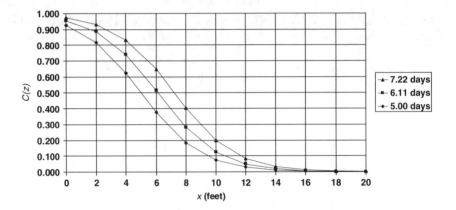

EXERCISE 2.7: Find the Taylor series and Maclaurin series for $f(z) = e^z$.

Solution: The Taylor series derivatives are

$$f^{(0)}(a) = e^a, \quad \frac{df(z)}{dz} = \frac{de^z}{dz} = e^z$$

$$f^{(m)}(a) = \frac{d^m f(z)}{dz^m}\bigg|_{z=a} = e^z|_{z=a} = e^a$$

Substitute the derivative terms into the Taylor series to get

$$f(z) = \sum_{m=0}^{\infty} \frac{f^{(m)}(a)}{m!}(z-a)^m$$

$$= e^a + \frac{e^a}{1!}(z-a) + \frac{e^a}{2!}(z-a)^2 + \cdots$$

$$= e^a\left[1 + (z-a) + \frac{(z-a)^2}{2!} + \cdots\right]$$

The Maclaurin series corresponds to $a = 0$; thus from the Taylor series we have

$$e^z = 1 + z + \frac{z^2}{2!} + \cdots$$

EXERCISE 2.8: Keep only first-order terms in z for small z and estimate asymptotic functional forms for $(1-z)^{-1}$, e^z, $\cos z$, $\sin z$, $\cosh z$, $\sinh z$, and $\ln(1+z)$.

Solution: The series representation of each function is used to find the following asymptotic approximations:

$$\frac{1}{1-z} \approx 1+z$$

$$e^z \approx 1+z$$

$$\cos z \approx 1 - \frac{z^2}{2} \approx 1 \quad \text{(drop second-order term)}$$

$$\sin z \approx z$$

$$\cosh z \approx 1 + \frac{z^2}{2} \approx 1 \quad \text{(drop second-order term)}$$

$$\sinh z \approx z$$

$$\ln(1+z) \approx z$$

S3 ANALYTIC GEOMETRY

EXERCISE 3.1: Fit the quadratic equation $y = a + bx + cx^2$ to the three points (x_1, y_1), (x_2, y_2), (x_3, y_3).

Solution: We must simultaneously solve three equations for the three unknown constants a, b, c. The equations to be solved are

$$y_1 = a + bx_1 + cx_1^2 \tag{i}$$

$$y_2 = a + bx_2 + cx_2^2 \tag{ii}$$

$$y_3 = a + bx_3 + cx_3^2 \tag{iii}$$

Rearrange Eq. (i) to find a in terms of b and c:

$$a = y_1 - bx_1 - cx_1^2 \tag{iv}$$

Subtract Eq. (i) from Eq. (ii) to eliminate a:

$$y_2 - y_1 = b(x_2 - x_1) + c(x_2^2 - x_1^2) \tag{v}$$

Solve Eq. (v) for b in terms of c and simplify:

$$b = \frac{y_2 - y_1}{x_2 - x_1} - c\frac{x_2^2 - x_1^2}{x_2 - x_1} = \frac{y_2 - y_1}{x_2 - x_1} - c(x_1 + x_2) \tag{vi}$$

Subtract Eq. (iii) from Eq. (i) and use Eq. (vi) to eliminate b:

$$y_3 - y_1 = b(x_3 - x_1) + c(x_3^2 - x_1^2)$$
$$= \left[\frac{y_2 - y_1}{x_2 - x_1} - c(x_1 + x_2) + c(x_1 + x_3)\right](x_3 - x_1) \tag{vii}$$

Solve Eq. (vii) for c:

$$c = \frac{1}{(x_3 - x_2)}\left[\frac{y_3 - y_1}{x_3 - x_1} - \frac{y_2 - y_1}{x_2 - x_1}\right] \tag{viii}$$

Find a, b, c by using Eq. (viii), then Eq. (vi) and, finally, Eq. (iv).

EXERCISE 3.2: Suppose $w = \alpha u^\beta$, where α, β are constants and u, w are variables. Take the logarithm of the equation and fit a least squares regression line to the resulting linear equation.

Solution: Rather than performing regression analysis on the original equation, a simpler way to approach the problem is to transform the equation to a new set of variables. If we take the logarithm of the equation, we have

$$\log w = \log \alpha + \beta \log u$$

By making the identifications

$$y = \log w$$
$$x = \log u$$
$$m = \beta$$
$$b = \log \alpha$$

we express the original equation in the form of a straight line

$$y = mx + b$$

and our previous regression analysis applies.

EXERCISE 3.3: Express $z = z_1 z_2$ and $z = z_1/z_2$ in polar coordinates.

Solution: Define the polar coordinate forms of z_1 and z_2 as

$$z_1 = r_1 e^{i\theta_1}, z_2 = r_2 e^{i\theta_2}$$

so that the product of z_1 and z_2 is

$$z = z_1 z_2 = r_1 r_2 e^{i(\theta_1 + \theta_2)}$$

Therefore the magnitude of z is

$$|z| = r_1 r_2$$

And the argument of z is

$$\arg(z) = \theta_1 + \theta_2 = \arg(z_1) + \arg(z_2)$$

up to multiples of 2π.

Using these definitions we write the ratio of z_1 and z_2 as

$$z = \frac{z_1}{z_2} = \frac{r_1}{r_2} e^{i(\theta_1 - \theta_2)}$$

Therefore the magnitude of z is

$$|z| = \frac{r_1}{r_2}$$

and the argument of z is

$$\arg(z) = \theta_1 - \theta_2 - \arg(z_1) = \arg(z_2)$$

S4   LINEAR ALGEBRA I

EXERCISE 4.1: The Pauli matrices are the square matrices

$$\sigma_x = \begin{bmatrix} 0 & 1 \\ 1 & 0 \end{bmatrix}, \sigma_y = \begin{bmatrix} 0 & -i \\ i & 0 \end{bmatrix}, \sigma_z = \begin{bmatrix} 1 & 0 \\ 0 & -1 \end{bmatrix}$$

These matrices are used to calculate the spin angular momentum of particles such as electrons, protons, and neutrons. Show that the Pauli matrices are Hermitian.

Solution: The matrix A is Hermitian if it is self-adjoint such that $A = A^+$; that is,

$$[a_{ij}]^+ = [a_{ij}^*]^T = [a_{ji}^*]$$

Find the conjugate transpose of the Pauli matrices:

$$\sigma_x^+ = \begin{bmatrix} 0 & 1 \\ 1 & 0 \end{bmatrix} = \sigma_x$$

$$\sigma_y^+ = [\sigma_y^*]^T = \begin{bmatrix} 0 & i \\ -i & 0 \end{bmatrix}^T = \begin{bmatrix} 0 & -i \\ i & 0 \end{bmatrix} = \sigma_y$$

$$\sigma_z^+ = [\sigma_z^*]^T = \begin{bmatrix} 1 & 0 \\ 0 & -1 \end{bmatrix} = \sigma_z$$

The Pauli matrices are self-adjoint, therefore they are Hermitian.

EXERCISE 4.2: Calculate

$$\begin{bmatrix} 3 & 4 & 2 \\ 2 & 3 & -1 \end{bmatrix} \begin{bmatrix} 1 & -2 & -4 \\ 0 & -1 & 2 \\ 6 & -3 & 9 \end{bmatrix}$$

Solution

$$\begin{bmatrix} 3 & 4 & 2 \\ 2 & 3 & -1 \end{bmatrix} \begin{bmatrix} 1 & -2 & -4 \\ 0 & -1 & 2 \\ 6 & -3 & 9 \end{bmatrix}$$

$$= \begin{bmatrix} 3 \cdot 1 + 4 \cdot 0 + 2 \cdot 6 & 3(-2) + 4(-1) + 2(-3) & 3(-4) + 4 \cdot 2 + 2 \cdot 9 \\ 2 \cdot 1 + 3 \cdot 0 + (-1)6 & 2(-2) + 3(-1) + (-1)(-3) & 2(-4) + 3 \cdot 2 + (-1)9 \end{bmatrix}$$

$$= \begin{bmatrix} 15 & -16 & 14 \\ -4 & -4 & -11 \end{bmatrix}$$

EXERCISE 4.3: Calculate

$$\begin{bmatrix} 2 & 3 & -1 \end{bmatrix} \begin{bmatrix} -4 \\ 2 \\ 9 \end{bmatrix}$$

Solution

$$\begin{bmatrix} 2 & 3 & -1 \end{bmatrix} \begin{bmatrix} -4 \\ 2 \\ 9 \end{bmatrix} = 2 \cdot (-4) + 3 \cdot 2 + (-1) \cdot 9 = -11$$

EXERCISE 4.4: Suppose we are given the Pauli matrices

$$\sigma_1 = \begin{bmatrix} 0 & 1 \\ 1 & 0 \end{bmatrix}, \sigma_2 = \begin{bmatrix} 0 & -i \\ i & 0 \end{bmatrix}, \sigma_3 = \begin{bmatrix} 1 & 0 \\ 0 & -1 \end{bmatrix}$$

(a) Calculate σ_3^2. (b) Calculate the trace of σ_1 and the trace of σ_3. (c) Verify the equality $[\sigma_2 \sigma_3]^T = \sigma_3^T \sigma_2^T$.

Solution: (a) The square of σ_3 is

$$\sigma_3^2 = \begin{bmatrix} 1 & 0 \\ 0 & -1 \end{bmatrix} \begin{bmatrix} 1 & 0 \\ 0 & -1 \end{bmatrix} = \begin{bmatrix} 1 & 0 \\ 0 & 1 \end{bmatrix} = I$$

(b) The trace of σ_i equals 0 for all $i = 1, 2, 3$. For example, $\text{Tr}(\sigma_3) = (\sigma_3)_{11} + (\sigma_3)_{22} = 1 + (-1) = 0$.

(c) Calculate the left-hand side

$$[\boldsymbol{\sigma}_2 \ \boldsymbol{\sigma}_3]^T = \left\{ \begin{bmatrix} 0 & -i \\ i & 0 \end{bmatrix} \begin{bmatrix} 1 & 0 \\ 0 & -1 \end{bmatrix} \right\}^T = \begin{bmatrix} 0 & i \\ i & 0 \end{bmatrix}^T = \begin{bmatrix} 0 & i \\ i & 0 \end{bmatrix}$$

and right-hand side

$$\boldsymbol{\sigma}_3^T \boldsymbol{\sigma}_2^T = \begin{bmatrix} 1 & 0 \\ 0 & -1 \end{bmatrix} \begin{bmatrix} 0 & i \\ -i & 0 \end{bmatrix} = \begin{bmatrix} 0 & i \\ i & 0 \end{bmatrix}$$

Comparing results verifies that $[\boldsymbol{\sigma}_2 \ \boldsymbol{\sigma}_3]^T = \boldsymbol{\sigma}_3^T \ \boldsymbol{\sigma}_2^T$.

EXERCISE 4.5: Let

$$A = \begin{bmatrix} 2 & 5 \\ 1 & 3 \end{bmatrix}, A^{-1} = \begin{bmatrix} 3 & -5 \\ -1 & 2 \end{bmatrix}$$

Verify $AA^{-1} = I$.

Solution:

$$AA^{-1} = \begin{bmatrix} 2 & 5 \\ 1 & 3 \end{bmatrix} \begin{bmatrix} 3 & -5 \\ -1 & 2 \end{bmatrix} = \begin{bmatrix} 2 \cdot 3 + 5(-1) & 2(-5) + 5 \cdot 2 \\ 1 \cdot 3 + 3(-1) & 1(-5) + 3 \cdot 2 \end{bmatrix} = \begin{bmatrix} 1 & 0 \\ 0 & 1 \end{bmatrix}$$

EXERCISE 4.6: Let

$$A = \begin{bmatrix} 1 & 0 & 2 \\ 2 & -1 & 3 \\ 4 & 1 & 8 \end{bmatrix}, A^{-1} = \begin{bmatrix} -11 & 2 & 2 \\ -4 & 0 & 1 \\ 6 & -1 & -1 \end{bmatrix}$$

Verify $AA^{-1} = I$.

Solution:

$$AA^{-1} = \begin{bmatrix} 1 & 0 & 2 \\ 2 & -1 & 3 \\ 4 & 1 & 8 \end{bmatrix} \begin{bmatrix} -11 & 2 & 2 \\ -4 & 0 & 1 \\ 6 & -1 & -1 \end{bmatrix}$$

$$= \begin{bmatrix} 1(-11) + 0(-4) + 2 \cdot 6 & 1 \cdot 2 + 0 \cdot 0 + 2(-1) & 1 \cdot 2 + 0 \cdot 1 + 2(-1) \\ 2(-11) - 1(-4) + 3 \cdot 6 & 2 \cdot 2 - 1 \cdot 0 + 3(-1) & 2 \cdot 2 - 1 \cdot 1 + 3(-1) \\ 4(-11) + 1(-4) + 8 \cdot 6 & 4 \cdot 2 + 1 \cdot 0 + 8(-1) & 4 \cdot 2 + 1 \cdot 1 + 8(-1) \end{bmatrix}$$

$$= \begin{bmatrix} 1 & 0 & 0 \\ 0 & 1 & 0 \\ 0 & 0 & 1 \end{bmatrix} = I$$

EXERCISE 4.7: Let

$$A = \begin{bmatrix} a & b \\ c & d \end{bmatrix}, A^{-1} = \begin{bmatrix} x & y \\ z & w \end{bmatrix}$$

Assume a, b, c, d are known. Find x, y, z, w from $AA^{-1} = I$.

Solution:

$$AA^{-1} = \begin{bmatrix} a & b \\ c & d \end{bmatrix}\begin{bmatrix} x & y \\ z & w \end{bmatrix} = \begin{bmatrix} ax + bz & ay + bw \\ cx + dz & cy + bw \end{bmatrix}$$

Note that $\{a, b, c, d\}$ are known. To find $\{x, y, z, w\}$ we recognize that $I = \begin{bmatrix} 1 & 0 \\ 0 & 1 \end{bmatrix}$. Equating AA^{-1} and I gives us four equations in four unknowns; thus

$$ax + bz = 1, \, ay + bw = 0$$
$$cx + dz = 0, \, cy + dw = 1$$

Solving for x, y, z, w gives

$$x = \frac{d}{|A|}, \, y = \frac{-b}{|A|}, \, z = \frac{-c}{|A|}, \, w = \frac{a}{|A|}$$

where $|A| = ad - bc$.

EXERCISE 4.8: Assume b_1, b_2 are known constants. Write the system of equations

$$2x_1 + 5x_2 = b_1$$
$$x_1 + 3x_2 = b_2$$

in matrix form and solve for the unknowns x_1, x_2.

Solution: We want to write the system of equations as $Ax = b$. In this case we have

$$A = \begin{bmatrix} 2 & 5 \\ 1 & 3 \end{bmatrix}, x = \begin{bmatrix} x_1 \\ x_2 \end{bmatrix}, b = \begin{bmatrix} b_1 \\ b_2 \end{bmatrix} \tag{i}$$

The solution is $x = A^{-1}b$. We find the inverse of matrix A using the equation $A^{-1}A = I$, where I is the identity matrix; thus

$$A^{-1}A = \begin{bmatrix} \alpha & \beta \\ \gamma & \delta \end{bmatrix}\begin{bmatrix} 2 & 5 \\ 1 & 3 \end{bmatrix} = \begin{bmatrix} 1 & 0 \\ 0 & 1 \end{bmatrix} = I \tag{ii}$$

This gives four equations for the four unknown elements α, β, γ, δ of A^{-1}; namely,

$$2\alpha + \beta = 1$$
$$5\alpha + 3\beta = 0$$
$$2\gamma + \delta = 0 \tag{iii}$$
$$5\gamma + 3\delta = 1$$

These equations give $\alpha = 3$, $\beta = -5$, $\gamma = -1$, $\delta = 2$. These values can be checked by substituting into Eq. (iii) and verifying the equalities. The inverse matrix becomes

$$A^{-1} = \begin{bmatrix} \alpha & \beta \\ \gamma & \delta \end{bmatrix} = \begin{bmatrix} 3 & -5 \\ -1 & 2 \end{bmatrix} \tag{iv}$$

The relationship $A^{-1}A = I$ is verified in Exercise 4.5. Use Eq. (iv) in $x = A^{-1}b$ to find the unknowns x_1, x_2; thus

$$\begin{bmatrix} x_1 \\ x_2 \end{bmatrix} = \begin{bmatrix} 3 & -5 \\ -1 & 2 \end{bmatrix} \begin{bmatrix} b_1 \\ b_2 \end{bmatrix} = \begin{bmatrix} 3b_1 - 5b_2 \\ -b_1 + 2b_2 \end{bmatrix} \tag{v}$$

EXERCISE 4.9: Evaluate

$$\det \begin{bmatrix} 2 & 4 & 3 \\ 6 & 1 & 5 \\ -2 & 1 & 3 \end{bmatrix}$$

Solution:

$$\det \begin{bmatrix} 2 & 4 & 3 \\ 6 & 1 & 5 \\ -2 & 1 & 3 \end{bmatrix}$$
$$= 2 \cdot 1 \cdot 3 + 4 \cdot 5(-2) + 3 \cdot 6 \cdot 1 - (-2) \cdot 1 \cdot 3 - 1 \cdot 5 \cdot 2 - 3 \cdot 6 \cdot 4$$
$$= 6 - 40 + 18 - (-6) - 10 - 72$$
$$= -92$$

EXERCISE 4.10: Use the definition of cofactor to evaluate

$$\operatorname{cof}_{23} \begin{bmatrix} 2 & 4 & 3 \\ 6 & 1 & 5 \\ -2 & 1 & 3 \end{bmatrix}$$

Solution:

$$\text{cof}_{23} \begin{bmatrix} 2 & 4 & 3 \\ 6 & 1 & 5 \\ -2 & 1 & 3 \end{bmatrix} = - \begin{vmatrix} 2 & 4 \\ -2 & 1 \end{vmatrix} = -10$$

EXERCISE 4.11: Evaluate the following determinant using cofactors:

$$|A| = \det \begin{bmatrix} 2 & 4 & 3 \\ 6 & 1 & 5 \\ -2 & 1 & 3 \end{bmatrix}$$

Solution: Expand the determinant in cofactors:

$$|A| = \begin{vmatrix} 2 & 4 & 3 \\ 6 & 1 & 5 \\ -2 & 1 & 3 \end{vmatrix} = a_{21} \, \text{cof}_{21} + a_{22} \, \text{cof}_{22} + a_{23} \, \text{cof}_{23}$$

Evaluating cofactors gives

$$\text{cof}_{23} = -10$$

from Exercise 4.10, and

$$\text{cof}_{21} = - \begin{vmatrix} 4 & 3 \\ 1 & 3 \end{vmatrix} = -9$$

$$\text{cof}_{22} = + \begin{vmatrix} 2 & 3 \\ -2 & 3 \end{vmatrix} = 12$$

Substituting the cofactors into the expression for the determinant gives

$$\begin{aligned} |A| &= 6(-9) + 1(12) + 5(-10) \\ &= -54 + 12 - 50 \\ &= -92 \end{aligned}$$

in agreement with Exercise 4.9.

EXERCISE 4.12: Use Cramer's rule to find the unknowns $\{x_1, x_2, x_3\}$ that satisfy the system of equations

$$\begin{aligned} x_1 + 2x_3 &= 11 \\ 2x_1 - x_2 + 3x_3 &= 17 \\ 4x_1 + x_2 + 8x_3 &= 45 \end{aligned} \tag{i}$$

Solution: Write Eq. (i) in matrix form $Ax = b$, where

$$A = \begin{bmatrix} 1 & 0 & 2 \\ 2 & -1 & 3 \\ 4 & 1 & 8 \end{bmatrix} \tag{ii}$$

and

$$b = \begin{bmatrix} 11 \\ 17 \\ 45 \end{bmatrix} \tag{iii}$$

The determinant of A is given by

$$\det A = \det \begin{bmatrix} 1 & 0 & 2 \\ 2 & -1 & 3 \\ 4 & 1 & 8 \end{bmatrix} \tag{iv}$$

$$= 1 \cdot (-1) \cdot 8 + 0 \cdot 3 \cdot 4 + 2 \cdot 2 \cdot 1 - 4 \cdot (-1) \cdot 2 - 1 \cdot 3 \cdot 1 - 8 \cdot 2 \cdot 0$$
$$= -8 + 4 + 8 - 3 = 1$$

The determinant of A_j is found by replacing column j with b; thus

$$\det A_1 = \det \begin{bmatrix} 11 & 0 & 2 \\ 17 & -1 & 3 \\ 45 & 1 & 8 \end{bmatrix} = 3 \tag{v}$$

$$\det A_2 = \det \begin{bmatrix} 1 & 11 & 2 \\ 2 & 17 & 3 \\ 4 & 45 & 8 \end{bmatrix} = 1 \tag{vi}$$

and

$$\det A_3 = \det \begin{bmatrix} 1 & 0 & 11 \\ 2 & -1 & 17 \\ 4 & 1 & 45 \end{bmatrix} = 4 \tag{vii}$$

The unknowns are found from Cramer's rule as

$$x_1 = \frac{\det A_1}{\det A} = 3, \; x_2 = \frac{\det A_2}{\det A} = 1, \; x_3 \frac{\det A_3}{\det A} = 4 \tag{viii}$$

As a check of our solution, we note that the matrix equation $Ax = b$ has the formal solution $x = A^{-1}b$. The inverse of matrix A is given in Exercise 4.6. Carrying out the matrix multiplication

$$x = A^{-1}b = \begin{bmatrix} -11 & 2 & 2 \\ -4 & 0 & 1 \\ 6 & -1 & -1 \end{bmatrix} \begin{bmatrix} 11 \\ 17 \\ 45 \end{bmatrix} = \begin{bmatrix} 3 \\ 1 \\ 4 \end{bmatrix} \tag{ix}$$

which verifies the solution found by Cramer's rule.

S5   LINEAR ALGEBRA II

EXERCISE 5.1: Define $C = B - A$. Derive the law of cosines by calculating $C \cdot C$ for $\theta \leq 90°$.

Solution: Observe that we may express C in terms of the angle θ, which is the angle between A and B. Now

$$\begin{aligned} |C|^2 = C \cdot C &= (B - A) \cdot (B - A) \\ &= (B \cdot B - A \cdot B - B \cdot A + A \cdot A) \\ &= |B|^2 + |A|^2 - 2|A|\,|B|\cos\theta \end{aligned} \tag{i}$$

where we have used the commutative property of the dot product and its definition in terms of the angle θ. Equation (i) is the law of cosines.

EXERCISE 5.2: Let

$$A = \begin{bmatrix} a_{11} & a_{12} \\ a_{21} & a_{22} \end{bmatrix}$$

Find the characteristic equation and the characteristic roots.

Solution: The characteristic equation is

$$\begin{aligned} |A - \lambda I| = \begin{vmatrix} a_{11} - \lambda & a_{12} \\ a_{21} & a_{22} - \lambda \end{vmatrix} &= (a_{11} - \lambda)(a_{22} - \lambda) - a_{21}a_{12} \\ &= a_{11}a_{22} - \lambda(a_{22} + a_{11}) + \lambda^2 - a_{21}a_{12} \\ &= \lambda^2 - \lambda(a_{11} + a_{22}) + (a_{11}a_{22} - a_{21}a_{12}) = 0 \end{aligned}$$

Solution of the characteristic equation gives the characteristic roots. In this case, the quadratic formula gives

$$\lambda_{\pm} = \frac{(a_{11} + a_{22})}{2} \pm \frac{\sqrt{(a_{11} + a_{22})^2 - 4(a_{11}a_{22} - a_{21}a_{12})}}{2}$$

EXERCISE 5.3: Find eigenvalues for the Pauli matrices given in Exercise 4.1; that is, find λ from

$$\sigma_i \psi = \lambda I \psi$$

where ψ is an eigenfunction and $i = \{x, y, z\}$.

Solution: The eigenvalues are found by solving the equation

$$|\sigma_i - \lambda I| = 0$$

for $i = 1, 2, 3$.
   (a) $i = 1$:

$$\begin{vmatrix} \lambda & 1 \\ 1 & \lambda \end{vmatrix} = \lambda^2 - 1 = 0$$

implies the eigenvalues $\lambda_{\pm} = \pm 1$.
   (b) $i = 2$:

$$\begin{vmatrix} \lambda & -i \\ i & \lambda \end{vmatrix} = \lambda^2 + i^2 = \lambda^2 - 1 = 0$$

implies $\lambda_{\pm} = \pm 1$.
   (c) $i = 3$:

$$\begin{vmatrix} 1 - \lambda & 0 \\ 0 & -1 - \lambda \end{vmatrix} = -(1 - \lambda)(1 + \lambda) = -(1 - \lambda^2) = \lambda^2 - 1 = 0$$

implies $\lambda_{\pm} = \pm 1$. The eigenvalues of each Pauli matrix are ± 1.

EXERCISE 5.4: Use the Hamilton–Cayley theorem to find the inverse of the complex Pauli matrix

$$\sigma_y = \begin{bmatrix} 0 & -i \\ i & 0 \end{bmatrix}$$

Solution: The characteristic equation is

$$\begin{vmatrix} 0 - \lambda & -i \\ i & 0 - \lambda \end{vmatrix} = \lambda^2 - 1 = 0$$

so that the coefficients of the polynomial $f(\boldsymbol{\sigma}_y)$ for the $n = 2$ matrix $\boldsymbol{\sigma}_y$ are

$$a_0 = 1,\ a_1 = 0,\ a_2 = -1$$

By the Hamilton–Cayley theorem

$$\boldsymbol{\sigma}_y^{-1} = -\frac{1}{a_2}(a_0\boldsymbol{\sigma}_y + a_1)$$

Substituting in values of the coefficients gives

$$\boldsymbol{\sigma}_y^{-1} = -\frac{1}{(-1)}(1 \cdot \boldsymbol{\sigma}_y + 0) = \boldsymbol{\sigma}_y$$

In other words, the Pauli matrix $\boldsymbol{\sigma}_y$ is its own inverse, which is easily verified.

EXERCISE 5.5: Calculate the eigenvalues of the matrix $\boldsymbol{D}$.

Solution: The eigenvalue problem is $\det[\boldsymbol{D} - \lambda\boldsymbol{I}] = 0$. Using the expression for $\boldsymbol{D}$ gives

$$\begin{vmatrix} d_{11} - \lambda & \varepsilon_{12} \\ \varepsilon_{21} & d_{22} - \lambda \end{vmatrix} = 0$$

or

$$(d_{11} - \lambda)(d_{22} - \lambda) - \varepsilon_{12}\varepsilon_{21} = 0$$

Recall that $\varepsilon_{12} = \varepsilon_{21}$ then expand the characteristic equation to get

$$d_{11}d_{22} - \lambda(d_{11} + d_{22}) + \lambda^2 - \varepsilon_{12}^2 = 0$$

The eigenvalues are found from the quadratic equation to be

$$\lambda_\pm = \tfrac{1}{2}(d_{11} + d_{22}) \pm \tfrac{1}{2}[(d_{11} + d_{22})^2 - 4(d_{11}d_{22} - \varepsilon_{12}^2)]^{1/2}$$

EXERCISE 5.6: Show that $\boldsymbol{a}^+$ and $\boldsymbol{a}^-$ are orthogonal.

Solution: To show that $\boldsymbol{a}^+$ and $\boldsymbol{a}^-$ are orthogonal, we must show that

$$\boldsymbol{a}^+ \cdot \boldsymbol{a}^- = a_1^+ a_1^- + a_2^+ a_2^- = 0$$

Substituting in the expressions for a^+, a^- and simplifying the algebra gives

$$\frac{-\varepsilon_{12}}{d_{11} - \lambda_+} \left\{ 1 + \frac{\varepsilon_{12}^2}{(d_{11} - \lambda_+)^2} \right\}^{-1/2} \cdot \frac{-(d_{11} - \lambda_+)}{[(d_{11} - \lambda_+)^2 + \varepsilon_{12}^2]^{1/2}}$$

$$+ \left\{ 1 + \frac{\varepsilon_{12}^2}{(d_{11} - \lambda_+)^2} \right\}^{-1/2} \cdot \frac{-\varepsilon_{12}}{[(d_{11} - \lambda_+)^2 + \varepsilon_{12}^2]^{1/2}}$$

$$= (-\varepsilon_{12}) \cdot \frac{-(d_{11} - \lambda_+)}{(d_{11} - \lambda_+)^2 + \varepsilon_{12}^2} + (d_{11} - \lambda_+) \cdot \frac{-\varepsilon_{12}}{(d_{11} - \lambda_+)^2 + \varepsilon_{12}^2} = 0$$

as expected.

EXERCISE 5.7: Verify that $a_1^+ = a_2^-$ and $a_2^+ = -a_1^-$. Find θ.

Solution: Let $B = d_{11} - \lambda_+$; thus

$$a_1^+ = -\frac{\varepsilon_{12}}{B} \left[1 + \frac{\varepsilon_{12}^2}{B^2} \right]^{-1/2} = -\frac{\varepsilon_{12}}{B} \left[\frac{B^2 + \varepsilon_{12}^2}{B^2} \right]^{-1/2}$$

Further simplifying yields

$$a_1^+ = -\frac{\varepsilon_{12}}{B} \left[\frac{B^2}{B^2 + \varepsilon_{12}^2} \right]^{1/2} = -\frac{\varepsilon_{12}}{B} \frac{B}{[B^2 + \varepsilon_{12}^2]^{1/2}} = -\frac{\varepsilon_{12}}{[B^2 + \varepsilon_{12}^2]^{1/2}} = a_2^-$$

A similar calculation for a_2^+ gives

$$a_2^+ = \left[1 + \frac{\varepsilon_{12}^2}{B^2} \right]^{-1/2} = \left[\frac{B^2 + \varepsilon_{12}^2}{B^2} \right]^{-1/2} = \frac{B}{[B^2 + \varepsilon_{12}^2]^{1/2}} = -a_1^-$$

The angle θ may be found from either

$$\theta = \cos^{-1} a_1^+$$

or

$$\theta = \sin^{-1} a_1^-$$

Note that

$$\cos^2 \theta + \sin^2 \theta = (a_1^+)^2 + (-a_2^+)^2 = (a_1^-)^2 + (a_2^-)^2 = 1$$

which demonstrates that the eigenvectors are orthonormal.

S6   DIFFERENTIAL CALCULUS

EXERCISE 6.1: Let $f(x) = 3x + 7$, $g(x) = 5x + 1$, so that $f(x) + g(x) = 8x + 8$. Find $\lim_{x \to 2} [f(x) + g(x)]$.

Solution: Evaluate limits of functions separately.

$$\lim_{x \to 2} f(x) = 13 \text{ and } \lim_{x \to 2} g(x) = 11$$

Also

$$\lim_{x \to 2} [f(x) + g(x)] = \lim_{x \to 2} [8x + 8] = 24$$

Therefore

$$\lim_{x \to 2} f(x) + \lim_{x \to 2} g(x) = 24 = \lim_{x \to 2} [f(x) + g(x)]$$

as expected.

EXERCISE 6.2: Let $f(x) = 3x + 7$ and $c = 6$. Find $\lim_{x \to 2} 6(3x + 7)$.

Solution:

$$\lim_{x \to 2} 6(3x + 7) = 6 \lim_{x \to 2} (3x + 7) = 6 \cdot 13 = 78$$

EXERCISE 6.3: Find

$$\lim_{x \to 2} \frac{3x + 7}{5x + 1}$$

Solution:

$$\lim_{x \to 2} \frac{3x + 7}{5x + 1} = \frac{\lim_{x \to 2} (3x + 7)}{\lim_{x \to 2} (5x + 1)} = \frac{13}{11}$$

EXERCISE 6.4: Let $P(x) = \alpha x^2 + \beta x + \gamma$. Find $\lim_{x \to 0} P(x)$.

Solution:

$$\lim_{x \to 0} P(x) = \lim_{x \to 0} [\alpha x^2 + \beta x + \gamma] = P(0) = \gamma$$

EXERCISE 6.5: Find

$$\lim_{x \to -1} \frac{x^3 + 5x^2 + 3x}{2x^5 + 3x^4 - 2x^2 - 1}$$

Solution:

$$\lim_{x \to -1} \frac{x^3 + 5x^2 + 3x}{2x^5 + 3x^4 - 2x^2 - 1} = \frac{\lim_{x \to -1} (x^3 + 5x^2 + 3x)}{\lim_{x \to -1} (2x^5 + 3x^4 - 2x^2 - 1)}$$

$$= \frac{(-1)^3 + 5(-1)^2 + 3(-1)}{2(-1)^5 + 3(-1)^4 - 2(-1)^2 - 1} = -\frac{1}{2}$$

Note that the limit also equals $N(-1)/D(-1)$, where $f(x) = N(x)/D(x)$ and we are evaluating $\lim_{x \to a} f(x)$.

EXERCISE 6.6: Use the definition of the derivative to find the derivative of each of the following functions: (a) $f(x) = ax$; (b) $f(x) = bx^2$; (c) $f(x) = e^x$; (d) $f(x) = \sin x$.

Solution: (a) Derivatives are found using the definition

$$\frac{df(x)}{dx} = \lim_{\Delta x \to 0} \frac{f(x + \Delta x) - f(x)}{\Delta x}$$

We first evaluate $f(x)$ and $f(x + \Delta x)$:

$$f(x) = ax, f(x + \Delta x) = a(x + \Delta x)$$

Now substitute these expressions into the definition of derivative:

$$\frac{df(x)}{dx} = \lim_{\Delta x \to 0} \frac{a(x + \Delta x) - ax}{\Delta x} = \lim_{\Delta x \to 0} \frac{a\Delta x}{\Delta x} = a$$

(b) Evaluate $f(x)$ and $f(x + \Delta x)$:

$$f(x) = bx^2$$

$$f(x + \Delta x) = b(x + \Delta x)^2 = b(x^2 + 2x\Delta x + \Delta x^2)$$

Evaluate $df(x)/dx$:

$$\frac{df(x)}{dx} = \lim_{\Delta x \to 0} \frac{b(x^2 + 2x\Delta x + \Delta x^2) - bx^2}{\Delta x}$$

$$= \lim_{\Delta x \to 0} \frac{b(2x\Delta x + \Delta x^2)}{\Delta x}$$

$$= \lim_{\Delta x \to 0} b(2x + \Delta x) = 2bx$$

(c) Evaluate $f(x)$ and $f(x + \Delta x)$:

$$f(x) = e^x, f(x + \Delta x) = e^x e^{\Delta x}$$

Evaluate $df(x)/dx$:

$$\frac{df(x)}{dx} = \lim_{\Delta x \to 0} \frac{e^x e^{\Delta x} - e^x}{\Delta x} = \lim_{\Delta x \to 0} \frac{e^x(e^{\Delta x} - 1)}{\Delta x}$$

From the series expansion of $e^{\Delta x}$ for Δx much less than 1 we get

$$e^{\Delta x} \approx 1 + \Delta x, \ \Delta x \ll 1$$

and

$$\frac{df(x)}{dx} = \lim_{\Delta x \to 0} \frac{e^x \Delta x}{\Delta x} = e^x$$

(d) Evaluate $f(x)$ and $f(x + \Delta x)$:

$$f(x) = \sin x$$
$$f(x + \Delta x) = \sin(x + \Delta x)$$
$$= \sin x \cos \Delta x + \sin \Delta x \cos x$$

where we have used the trigonometric identity $\sin(A + B) = \sin A \cos B + \sin B \cos A$. Now evaluate $df(x)/dx$:

$$\frac{df(x)}{dx} = \lim_{\Delta x \to 0} \frac{(\sin x \cos \Delta x + \sin \Delta x \cos x) - \sin x}{\Delta x}$$

As $\Delta x \to 0$, we have $\cos \Delta x \approx 1$ and $\sin \Delta x \approx \Delta x$. Therefore

$$\frac{df(x)}{dx} = \lim_{\Delta x \to 0} \frac{(\sin x + \Delta x \cos x) - \sin x}{\Delta x} = \cos x$$

EXERCISE 6.7: Evaluate

$$\lim_{x \to 0} \frac{x^2 + bx}{x}$$

where $g(x) = x^2 + bx$, and $f(x) = x$.

Solution: Note that $g(0) = 0$ and $f(0) = 0$. Therefore L'Hôpital's rule must be used; thus

$$\lim_{x \to 0} \frac{x^2 + bx}{x} = \lim_{x \to 0} \frac{2x + b}{1} = b$$

where $dg(x)/dx = 2x + b$ and $df(x)/dx = 1$. An alternative approach is to use factoring such that

$$\lim_{x \to 0} \frac{x^2 + bx}{x} = \lim_{x \to 0} (x + b) = b$$

EXERCISE 6.8: Evaluate

$$\lim_{x \to 0} \frac{x}{\ln^2 (1 + x)}$$

where $g(x) = x$, and $f(x) = \ln^2 (1 + x)$.

Solution: At $x = 0$ we have $g(0) = 0$ and $f(0) = 0$. To find the limit, we must use L'Hôpital's rule. First determine derivatives:

$$\frac{dg(x)}{dx} = 1; \quad \frac{df(x)}{dx} = \frac{2 \ln (1 + x)}{1 + x}$$

By L'Hôpital's rule, we have

$$\lim_{x \to 0} \frac{1}{\dfrac{2 \ln (1 + x)}{1 + x}} = \lim_{x \to 0} \frac{1 + x}{2 \ln (1 + x)}$$

In the limit as $x \to 0$ we have the asymptotic approximation $\ln (1 + x) \approx x$; thus

$$\lim_{x \to 0} \frac{g(x)}{f(x)} = \lim_{x \to 0} \frac{1 + x}{2x}$$

This limit has two solutions that depend on how $x \to 0$:

$$\lim_{x \to 0^+} \frac{1 + x}{2x} = \infty; \quad \lim_{x \to 0^+} \frac{1 + x}{2x} = -\infty$$

where $x \to 0^+$ implies x approaches 0 through positive values and $x \to 0^-$ implies x approaches 0 through negative values. Thus the function $x/\lfloor \ln^2 (1 + x) \rfloor$ is discontinuous at the point $x = 0$.

EXERCISE 6.9: Let $u = ax^2 + bx + c$. Find the second-order derivative.

Solution: Begin with the first-order derivative:

$$\frac{du}{dx} = \frac{d}{dx} (ax^2 + bx + c)$$

$$= \frac{d}{dx} (ax^2) + \frac{d}{dx} (bx) + \frac{d}{dx} c$$

$$= 2ax + b$$

Find the second-order derivative by differentiating the result of the first-order derivative:

$$\frac{d^2u}{dx^2} = \frac{d}{dx}\left(\frac{du}{dx}\right) = \frac{d}{dx}(2ax + b) = 2a$$

EXERCISE 6.10: (a) Calculate the differential of $y = x^3$. (b) Use the chain rule to calculate dy/dx, where x, y are given by the parametric equations

$$x = 2t + 3, \, y = t^2 - 1$$

Solution: (a) We calculate

$$dy = \frac{dy(x)}{dx} dx = 3x^2 dx$$

(b) By the chain rule,

$$\frac{dy}{dx} = \frac{dy}{dt}\frac{dt}{dx}$$

But

$$\frac{dt}{dx} = \frac{1}{dx/dt}$$

Therefore calculate

$$\frac{dy}{dx} = \frac{dy/dt}{dx/dt} = \frac{2t}{2} = t$$

Note that $t = (x - 3)/2$ from the parametric equation for x, so that

$$\frac{dy}{dx} = \frac{x - 3}{2}$$

An alternative approach is to write

$$y = t^2 - 1 = \left(\frac{x - 3}{2}\right)^2 - 1$$

Then we get the expected result

$$\frac{dy}{dx} = 2\left(\frac{x - 3}{2}\right)\frac{d}{dx}\left(\frac{x - 3}{2}\right) = (x - 3)\cdot\frac{1}{2} = \frac{x - 3}{2}$$

EXERCISE 6.11: Find the extrema of $y = ax^2 + bx + c$.

Solution: Evaluate the first and second derivatives of y:

$$y' = 2ax + b, \; y'' = 2a$$

Extrema are found by setting $y' = 0$ and finding corresponding values of x; thus

$$y' = 2ax_{\text{ext}} + b = 0$$

or

$$x_{\text{ext}} = -\frac{b}{2a}$$

From the second derivative of y, we know that $y'' = 2a > 0$ implies x_{ext} is a minimum and $y'' = 2a < 0$ implies x_{ext} is a maximum. For example, suppose $b = c = 0$ and $y = ax^2$. The extremum occurs at $x_{\text{ext}} = 0$. Sketches of $y = ax^2$ are shown as follows:

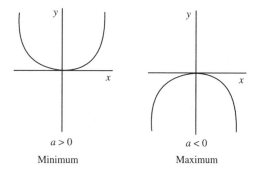

$a > 0$ $a < 0$

Minimum   Maximum

EXERCISE 6.12: Find the square root of any given positive number c. Show that the Newton–Raphson method works for $c = 2$.

Solution: Let $x = \sqrt{c}$ so that $c = x^2$ and $f(x) = x^2 - c = 0$. The derivative is $f'(x) = 2x$ and the iterative formula becomes

$$x_{n+1} = x_n - \frac{x_n^2 - c}{2x_n} = 0.5\left(x_n + \frac{c}{x_n}\right)$$

For $c = 2$ and $x_0 = 1$, we find

$$x_1 = 1.500\,000$$
$$x_2 = 1.416\,667$$
$$x_3 = 1.414\,216$$

$$\vdots$$

EXERCISE 6.13: Use the Taylor series to show that the centered difference $[P(x + \Delta x) - P(x - \Delta x)]/(2\Delta x)$ is second-order correct.

Solution: Use the Taylor series to write

$$P(x + \Delta x) = P(x) + \Delta x \frac{df}{dx} + \frac{\Delta x^2}{2} \frac{d^2 f}{dx^2} + \frac{\Delta x^3}{6} \frac{d^3 f}{dx^3} + \cdots$$

and

$$P(x - \Delta x) = P(x) - \Delta x \frac{df}{dx} + \frac{\Delta x^2}{2} \frac{d^2 f}{dx^2} - \frac{\Delta x^3}{6} \frac{d^3 f}{dx^3} + \cdots$$

Calculate the centered difference by first calculating

$$P(x + \Delta x) - P(x - \Delta x) = 2\Delta x \frac{df}{dx} + 2\frac{\Delta x^3}{6} \frac{d^3 f}{dx^3} + \cdots$$

Solve for df/dx:

$$\frac{df}{dx} = \frac{P(x + \Delta x) - P(x - \Delta x)}{2\Delta x} + E_T'$$

where the truncation error E_T' is

$$E_T' = -\frac{\Delta x^2}{6} \frac{d^3 f}{dx^3} + \cdots$$

The term E_T' is second order because of the term involving Δx^2. Neglecting terms second order and above means that E_T' is considered negligible, particularly in the limit as Δx approaches zero.

S7   PARTIAL DERIVATIVES

EXERCISE 7.1: Let $f(x, y) = e^{xy}$. Find f_{xy}, f_{yx}.

Solution: Given $f(x, y) = e^{xy}$, calculate

$$f_{xy} = \frac{\partial}{\partial y}\left(\frac{\partial f}{\partial x}\right) = \frac{\partial}{\partial y}(y e^{xy}) = (1 + xy) e^{xy}$$

$$f_{yx} = \frac{\partial}{\partial x}\left(\frac{\partial f}{\partial y}\right) = \frac{\partial}{\partial x}(x e^{xy}) = (1 + xy) e^{xy}$$

Note that $f_{xy} = f_{yx}$, as expected.

EXERCISE 7.2: Let $y = x_1^2 + x_1^2 x_2 + x_2^3$. Calculate $\partial y/\partial x_1$, $\partial y/\partial x_2$, and dy.

Solution: From the partial derivatives

$$\frac{\partial y}{\partial x_1} = 2x_1 + 2x_1 x_2, \quad \frac{\partial y}{\partial x_2} = x_1^2 + 3x_2^2$$

we can determine the total differential:

$$dy = \sum_{i=1}^{2} \frac{\partial y}{\partial x_i} dx_i = \frac{\partial y}{\partial x_1} dx_1 + \frac{\partial y}{\partial x_2} dx_2$$

$$= (2x_1 + 2x_1 x_2) dx_1 + (x_1^2 + 3x_2^2) dx_2$$

EXERCISE 7.3: Let $z = x^2 + y^2$, $x = at$, $y = be^{-t}$. Find dz/dt.

Solution: The total derivative is

$$\frac{dz}{dt} = \frac{\partial z}{\partial x}\frac{dx}{dt} + \frac{\partial z}{\partial y}\frac{dy}{dt} = 2x(a) + 2y(-be^{-t})$$

Replace x, y with their parametric representations to obtain

$$\frac{dz}{dt} = 2a^2 t - 2b^2 e^{-2t}$$

Alternatively, write z in terms of the parameter t only:

$$z = x^2 + y^2 = a^2 t^2 + b^2 e^{-2t}$$

Now evaluate the derivative:

$$\frac{dz}{dt} = 2a^2 t - 2b^2 e^{-2t}$$

which agrees with the total derivative calculated above.

EXERCISE 7.4: Find the Jacobian of the coordinate rotation $\mathbf{y} = \mathbf{a}\mathbf{x}$, where $\mathbf{a}$ is the matrix of coordinate rotations introduced in Chapter 4, Section 4.1.

Solution: The matrix a of the coordinate rotation is given by

$$a = \begin{bmatrix} \cos\theta & \sin\theta \\ -\sin\theta & \cos\theta \end{bmatrix} = [a_{ij}]$$

The Jacobian is found from

$$dy = J\,dx$$

where

$$J_{ij} = \frac{\partial y_i}{\partial x_j}$$

and

$$y_i = \sum_{j=1}^{2} a_{ij}x_j$$

Therefore

$$J_{ij} = \frac{\partial y_i}{\partial x_j} = a_{ij}$$

or $J = a$ for the two-dimensional coordinate rotation.

EXERCISE 7.5: Let $r = x\hat{i} + y\hat{j} + z\hat{k}$. Evaluate (a) $r = |r|$; (b) $\nabla \cdot r$; (c) $\nabla \times r$; (d) $\nabla(r)$; and (e) $\nabla(1/r)$.

Solution: (a) Magnitude of r:

$$r = |r| = \sqrt{r \cdot r} = \sqrt{x^2 + y^2 + z^2}$$

(b) Divergence of r:

$$\nabla \cdot r = \frac{\partial x}{\partial x} + \frac{\partial y}{\partial y} + \frac{\partial z}{\partial z} = 3$$

(c) Curl of r:

$$\nabla \times r = \begin{vmatrix} \hat{i} & \hat{j} & \hat{k} \\ \dfrac{\partial}{\partial x} & \dfrac{\partial}{\partial y} & \dfrac{\partial}{\partial z} \\ x & y & z \end{vmatrix}$$

$$= \hat{i}\begin{vmatrix} \dfrac{\partial}{\partial y} & \dfrac{\partial}{\partial z} \\ y & z \end{vmatrix} - \hat{j}\begin{vmatrix} \dfrac{\partial}{\partial x} & \dfrac{\partial}{\partial z} \\ x & z \end{vmatrix} + \hat{k}\begin{vmatrix} \dfrac{\partial}{\partial x} & \dfrac{\partial}{\partial y} \\ x & y \end{vmatrix}$$

Evaluating the derivatives resulting from expansion of the determinants shows that $\nabla \times \mathbf{r} = \mathbf{0}$.

(d) Gradient of $\mathbf{r}$:

$$\nabla r = \frac{\partial}{\partial x}(x^2 + y^2 + z^2)^{1/2}\,\hat{i} + \frac{\partial}{\partial y}(x^2 + y^2 + z^2)^{1/2}\,\hat{j}$$

$$+ \frac{\partial}{\partial z}(x^2 + y^2 + z^2)^{1/2}\hat{k}$$

$$= \frac{1}{2(x^2 + y^2 + z^2)^{1/2}}$$

$$\times \left\{ \frac{\partial}{\partial x}n(x^2 + y^2 + z^2)\hat{i} + \frac{\partial}{\partial y}(x^2 + y^2 + z^2)\hat{j} + \frac{\partial}{\partial z}(x^2 + y^2 + z^2)\hat{k} \right\}$$

$$= \frac{1}{2r}\{2x\hat{i} + 2y\hat{j} + 2z\hat{k}\}$$

$$= \frac{\mathbf{r}}{r}$$

(e) Gradient of $1/\mathbf{r}$:

$$\nabla\left(\frac{1}{r}\right) = \nabla(r^{-1}) = -\frac{1}{r^2}\nabla(r) = -\frac{\mathbf{r}}{r^3}$$

EXERCISE 7.6: Given $f(z) = z^2$, show that the Cauchy–Riemann equations are satisfied and evaluate $f'(z)$.

Solution: Expand $f(x)$ in terms of x, y:

$$f(z) = z^2$$
$$= x^2 - y^2 + 2ixy$$

Since $f(z) = u(x, y) + iv(x, y)$, we find

$$u(x, y) = x^2 - y^2$$
$$v(x, y) = 2xy$$

The Cauchy–Riemann equations give

$$\frac{\partial u}{\partial x} = 2x = \frac{\partial v}{\partial y}$$

$$-\frac{\partial u}{\partial y} = 2y = \frac{\partial v}{\partial x}$$

The derivative $f'(z)$ is

$$f'(z) = \frac{\partial u}{\partial x} + i\frac{\partial v}{\partial x} = 2x + i2y = 2(x + iy) = 2z$$

or

$$f'(z) = \frac{\partial v}{\partial y} - i\frac{\partial u}{\partial y} = 2x - i(-2y) = 2(x + iy) = 2z$$

S8 INTEGRAL CALCULUS

EXERCISE 8.1: Given

$$f(x) = \frac{1}{x}, x = g(t) = t^n$$

Find $\int_a^b f(x)\, dx$ by direct integration and by a change of variable.

Solution: Method 1 uses direct integration of $f(x)$ with respect to x:

$$\int_a^b f(x)\, dx = \int_a^b \frac{dx}{x} = \ln x\big|_a^b = \ln\frac{b}{a}$$

Method 2 involves performing integration with respect to t. First change the limits of integration:

$$g(\alpha) = a = \alpha^n, \, g(\beta) = b = \beta^n$$

This implies

$$\alpha = a^{1/n}, \beta = b^{1/n}$$

Change the integrand from a function of x to a function of t:

$$f(x)\, dx = f(g(t))\left[\frac{dg(t)}{dt}\right] dt = t^{-n}[nt^{n-1}]\, dt$$

where

$$f(x) = \frac{1}{x}$$

implies

$$f(g(t)) = \frac{1}{g(t)} = \frac{1}{t^n} = t^{-n}$$

Evaluate the definite integral:

$$\int_a^b f(x)\,dx = \int_\alpha^\beta f(g(t)) \left[\frac{dg(t)}{dt} \right] dt$$

$$= \int_{a^{1/n}}^{b^{1/n}} t^{-n} n t^{n-1}\,dt = n \int_{a^{1/n}}^{b^{1/n}} t^{-1}\,dt$$

$$= n \ln t \big|_{a^{1/n}}^{b^{1/n}} = n \ln \left(\frac{b}{a} \right)^{1/n} = \ln \left(\frac{b}{a} \right)$$

The results of Methods 1 and 2 agree, as expected.

EXERCISE 8.2: Evaluate the indefinite integral $I(x) = \int x \sin x\,dx$ using integration by parts.

Solution: Recall that integration by parts lets us write

$$\int u\,dv = uv - \int v\,du$$

Let $x = u$, $du = dx$, and $\sin x\,dx = dv$, so that

$$v = \int dv = \int \sin x\,dx = -\cos x$$

Then integration by parts gives

$$\int x \sin x\,dx = -x \cos x - \int (-\cos x)\,dx$$

$$= -x \cos x - (-\sin x) + C$$
$$= \sin x - x \cos x + C$$

where C is the integration constant.

The integral solution is verified by evaluating the derivative

$$\frac{d}{dx}(\sin x - x \cos x + C) = \frac{d \sin x}{dx} - \frac{dx \cos x}{dx} + \frac{dC}{dx}$$

$$= \cos x - [\cos x + x(-\sin x)] + 0$$
$$= x \sin x$$

EXERCISE 8.3: Calculate the length of a straight line $y = mx + b$ between the end points x_A, x_B.

Solution: For Method 1 the equation of a straight line is

$$y = mx + b$$

Therefore the slope of the line is

$$\frac{dy}{dx} = m$$

and the curve length L between the end points (x_A, x_B) is the integral

$$L = \int_{x_A}^{x_B} [1 + m^2]^{1/2} dx = [1 + m^2]^{1/2} \int_{x_A}^{x_B} dx$$
$$= [1 + m^2]^{1/2}(x_B - x_A)$$

For Method 2 the length L of a straight line from analytic geometry is

$$L = [(x_B - x_A)^2 + (y_B - y_A)^2]^{1/2}$$

where the points are sketched in the figure. Noting that the equation of a straight line lets us write

$$y_A = mx_A + b, \ y_B = mx_B + b$$

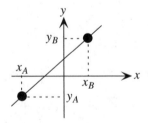

we find

$$y_B - y_A = (mx_B + b) - (mx_A + b) = m(x_B - x_A)$$

Substituting this result into L gives

$$L = [(x_B - x_A)^2 + m^2(x_B - x_A)^2]^{1/2} = [1 + m^2]^{1/2}(x_B - x_A)$$

in agreement with Method 1.

EXERCISE 8.4: Let $y = f(x) = x^2$. Evaluate the definite integral $\int_0^b f(x)\,dx$ analytically and numerically using the trapezoidal rule and Simpson's rule. Compare the results.

Solution: The exact solution is

$$\int_0^b x^2 dx = \frac{x^3}{3}\Big|_0^b = \frac{b^3}{3}$$

The trapezoidal rule gives

$$\int_0^b x^2 dx \approx \frac{h}{2} f(a) + \frac{h}{2} f(b) = \frac{b^3}{4}$$

since $h = b - 0 = b$ in this case. If we use Simpson's rule, we get

$$\int_0^b x^2 dx \approx \frac{h}{3} f(a) + \frac{4h}{3} f\left(\frac{a+b}{2}\right) + \frac{h}{3} f(b)$$

$$= 0 + \frac{4}{3}\left(\frac{b}{2}\right) f\left(\frac{b}{2}\right) + \frac{1}{3}\left(\frac{b}{2}\right) f(b)$$

where h is now $h = (b - a)/2 = b/2$. Thus

$$\int_0^b x^2 dx \approx \frac{4}{3}\left(\frac{b}{2}\right)\left(\frac{b}{2}\right)^2 + \frac{1}{3}\left(\frac{b}{2}\right) b^2 = \frac{b^3}{6} + \frac{b^3}{6} = \frac{b^3}{3}$$

In this case, Simpson's rule agrees exactly with the analytical solution.

S9 SPECIAL INTEGRALS

EXERCISE 9.1: Line Integral. Evaluate the integral $\int_c A \cdot dr$ from $(0, 0, 0)$ to $(1, 1, 1)$ along the path $x = t$, $y = t^2$, $z = t^3$ for

$$A = (3x^2 - 6yz)\hat{i} + (2y + 3xz)\hat{j} + (1 - 4xyz^2)\hat{k}$$

Solution: The line integral has the expanded form

$$\int_C A \cdot dr = \int_C \{(3x^2 - 6yz)\hat{i} + (2y + 3xz)\hat{j} + (1 - 4xyz^2)\hat{k}\}$$

$$\cdot [\hat{i}dx + \hat{j}dy + \hat{k}dz]$$

$$= \int_{(0,0,0)}^{(1,1,1)} \{(3x^2 - 6yz)\,dx + (2y + 3xz)\,dy + (1 - 4xyz^2)\,dz\}$$

The integral is evaluated by replacing x, y, z coordinates with their parameterized values. The points $(x, y, z) = (0, 0, 0)$ and $(1, 1, 1)$ correspond to parameter values $t = 0$ and 1, respectively; thus

$$\int_C A \cdot dr = \int_{t=0}^{1} \{[3t^2 - 6t^2(t^3)]\, dt + [2t^2 + 3t(t^3)]\, d(t^2)\}$$

$$+ \int_{t=0}^{1} [1 - 4t(t^2)(t^3)^2]\, d(t^3)$$

$$= \int_0^1 \{(3t^2 - 6t^5)\, dt + (4t^3 + 6t^5)\, dt + (3t^2 - 12t^{11})\, dt\}$$

$$= 2$$

EXERCISE 9.2: Double Integral. Evaluate the double integral $\int \int_R (x^2 + y^2)\, dx\, dy$ in the region R bounded by $y = x^2$, $x = 2$, $y = 1$.

Solution: The region R is shown in the sketch.

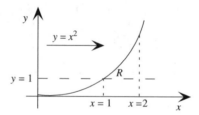

The double integral becomes

$$\int \int_R (x^2 + y^2)\, dx\, dy = \int_{x=1}^{2} \left\{ \int_{x=1}^{x^2} (x^2 + y^2)\, dy \right\} dx$$

The double integral is solved in this case by first performing the y integration and then the x integration; thus

$$\int \int_R (x^2 + y^2)\, dx\, dy = \int_{x=1}^{2} \left\{ \left[x^2 y + \frac{y^3}{3} \right]\Big|_{y=1}^{x^2} \right\} dx$$

$$= \int_{x=1}^{2} \left[x^4 + \frac{x^6}{3} - x^2 - \frac{1}{3} \right] dx$$

$$= \frac{1006}{105}$$

EXERCISE 9.3: Show that $\{\phi_n(x) = (\sin nx)/\sqrt{\pi}: n = 1, 2, 3, \ldots \}$ is an orthonormal set of functions in the interval $-\pi \leq x \leq \pi$.

Solution: To be orthonormal, we must show that

$$\frac{1}{\pi} \int_{-\pi}^{\pi} \sin mx \sin nx \, dx = \delta_{mn}$$

The integral is evaluated by using the trigonometric identities

$$\cos (A \pm B) = \cos A \cos B \mp \sin A \sin B$$

to derive the relation

$$2 \sin A \sin B = \cos (A - B) - \cos (A + B)$$

Substituting into the integral and solving for $m \neq n$ gives the orthogonality condition

$$\int_{-\pi}^{\pi} \sin mx \sin nx \, dx$$

$$= \frac{1}{2} \left\{ \int_{-\pi}^{\pi} [\cos (m - n)x] \, dx - \int_{-\pi}^{\pi} [\cos (m + n)x] \, dx \right\}$$

$$= \frac{1}{2} \left\{ \frac{\sin (m - n)x}{m - n} \Big|_{-\pi}^{\pi} - \frac{\sin (m + n)x}{m + n} \Big|_{-\pi}^{\pi} \right\} = 0 \text{ if } m \neq n$$

The normalization condition is found by solving the integral for $m = n$, to obtain

$$\frac{1}{\pi} \int_{-\pi}^{\pi} \sin nx \sin nx \, dx = \frac{1}{2\pi} \left\{ \int_{-\pi}^{\pi} dx - \int_{-\pi}^{\pi} [\cos 2nx] \, dx \right\}$$

$$= \frac{1}{2\pi} \left\{ x \Big|_{-\pi}^{\pi} - \frac{\sin 2nx}{2n} \Big|_{-\pi}^{\pi} \right\}$$

$$= 1$$

Notice that the constant $1/\sqrt{\pi}$ was needed in the definition of $\phi_n(x)$ to normalize the set of functions.

EXERCISE 9.4: Determine the Fourier transform of the harmonic function

$$g(t) = \begin{cases} \exp (i\omega_0 t), & -T \leq t \leq T \\ 0, & |t| > 0 \end{cases}$$

Solution: Calculate the integral

$$G(\omega) = \frac{1}{\sqrt{2\pi}} \int_{-T}^{T} e^{i\omega_0 t} e^{-i\omega t} \, dt = \frac{1}{\sqrt{2\pi}} \int_{-T}^{T} e^{i(\omega_0 - \omega)t} \, dt$$

Solve the integral to get

$$G(\omega) = \frac{1}{\sqrt{2\pi}} \frac{e^{i(\omega_0 - \omega)t}}{i(\omega_0 - \omega)} \Bigg|_{-T}^{T}$$

and simplify using Euler's equation:

$$G(\omega) = \sqrt{\frac{2}{\pi}} T \left[\frac{\sin(\omega_0 - \omega)T}{(\omega_0 - \omega)T} \right]$$

A plot of $G(\omega)$ versus ω shows that frequencies ω near ω_0 contribute most to the transform.

EXERCISE 9.5: Show that Eq. (9.4.8) is valid. *Hint*: You should use the Dirac delta function in Eq. (9.4.4).

Solution: Equation (9.4.8) is

$$f(t) = \frac{1}{(2\pi)} \int_{-\infty}^{\infty} \left[\int_{-\infty}^{\infty} f(s) e^{-i\omega s} ds \right] e^{i\omega t} d\omega \tag{i}$$

Rearranging gives

$$f(t) = \int_{-\infty}^{\infty} f(s) \left[\frac{1}{(2\pi)} \int_{-\infty}^{\infty} e^{i\omega(t-s)} d\omega \right] ds \tag{ii}$$

The bracketed term in Eq. (ii) is the Dirac delta function in Eq. (9.4.4), so Eq. (ii) becomes

$$f(t) = \int_{-\infty}^{\infty} f(s) \delta(t - s) ds \tag{iii}$$

Equation (iii) is equivalent to Eq. (9.4.3), which demonstrates that Eq. (9.4.8) is valid.

EXERCISE 9.6: Show that the Fourier transform of the product of two functions $g_1(t)$ and $g_2(t)$ can be written as a convolution integral. *Hint*: You should use Eqs. (9.4.11) and (9.4.12) and the Dirac delta function in Eq. (9.4.4).

Solution: Following Collins [1999], we begin with Eq. (9.4.11) to form the Fourier transform of the product of two functions $g_1(t)$ and $g_2(t)$; namely,

$$\mathcal{F}\{g_1(t)g_2(t)\} = \int_{-\infty}^{\infty} g_1(t)g_2(t) e^{-i\omega t} dt \tag{i}$$

The Fourier integral representations of the two functions $g_1(t)$ and $g_2(t)$ are

$$\mathcal{F}^{-1}\{B_1(\omega)\} = g_1(t) = \frac{1}{2\pi} \int_{-\infty}^{\infty} B_1(\omega)e^{i\omega t} d\omega \qquad \text{(ii)}$$

and

$$\mathcal{F}^{-1}\{B_2(\omega)\} = g_2(t) = \frac{1}{2\pi} \int_{-\infty}^{\infty} B_2(\omega)e^{i\omega t} d\omega \qquad \text{(iii)}$$

from Eq. (9.4.12). Substituting Eqs. (ii) and (iii) into Eq. (i) gives

$$\mathcal{F}\{g_1(t)g_2(t)\} = \frac{1}{4\pi^2} \int_{-\infty}^{\infty} \int_{-\infty}^{\infty} \int_{-\infty}^{\infty} B_1(\omega')B_2(\omega'')e^{-i(\omega-\omega'-\omega'')t} d\omega' d\omega'' dt \qquad \text{(iv)}$$

The integral over t is the Dirac delta function

$$\frac{1}{2\pi} \int_{-\infty}^{\infty} e^{-i(\omega-\omega'-\omega'')t} dt = \delta(\omega - \omega' - \omega'') \qquad \text{(v)}$$

Using Eq. (v) in Eq. (iv) lets us write

$$\mathcal{F}\{g_1(t)g_2(t)\} = \frac{1}{2\pi} \int_{-\infty}^{\infty} \int_{-\infty}^{\infty} B_1(\omega') B_2(\omega'') \delta(\omega - \omega' - \omega'') d\omega' d\omega'' \qquad \text{(vi)}$$

Two integrals are possible. The integral over ω' is

$$\mathcal{F}\{g_1(t)g_2(t)\} = \frac{1}{2\pi} \int_{-\infty}^{\infty} B_1(\omega - \omega'') B_2(\omega'') d\omega'' \qquad \text{(vii)}$$

The integral over ω'' is

$$\mathcal{F}\{g_1(t)g_2(t)\} = \frac{1}{2\pi} \int_{-\infty}^{\infty} B_1(\omega') B_2(\omega - \omega') d\omega' \qquad \text{(viii)}$$

Comparing Eqs. (vii) and (viii) with the form of the convolution integral in Eq. (9.4.14) shows that

$$\mathcal{F}\{g_1(t)g_2(t)\} = \frac{1}{2\pi} B_1(\omega - \omega'') \circ B_2(\omega'') = \frac{1}{2\pi} B_1(\omega') \circ B_2(\omega - \omega') \qquad \text{(ix)}$$

Thus the Fourier transform of the product of two functions $g_1(t)$ and $g_2(t)$ can be written as a convolution integral.

EXERCISE 9.7: What are the Z transforms of the causal digital functions $h_1 = \{1, -2, -1, 0, 1\}$ and $h_2 = \{0.6, -1.2, -2.4, 0.7, 3.8\}$?

Solution: Substitute the coefficients of each causal digital function into the Z transform to obtain $h_1(z) = 1 - 2z - z^2 + z^4$ and $h_2(z) = 0.6 - 1.2z - 2.4z^2 + 0.7z^3 + 3.8z^4$.

EXERCISE 9.8: Verify the Laplace transform $\mathcal{L}\{af(t)\} = a\mathcal{L}\{f(t)\}$, where a is a complex constant, and $f(t)$ is a function of the variable t.

Solution: Use the definition of Laplace transform to write

$$\mathcal{L}\{af(t)\} = \int_0^\infty e^{-st} af(t)\, dt = a \int_0^\infty e^{-st} f(t)\, dt = a\mathcal{L}\{f(t)\}$$

EXERCISE 9.9: Verify the inverse Laplace transform $\mathcal{L}^{-1}\{aF(s)\} = a\mathcal{L}^{-1}\{F(s)\}$, where a is a complex constant, and $F(s) = \mathcal{L}\{f(t)\}$ is the Laplace transform of a function $f(t)$ of the variable t.

Solution: Note from Exercise 9.8 that $\mathcal{L}\{af(t)\} = a\mathcal{L}\{f(t)\}$. We can therefore write

$$\mathcal{L}^{-1}\{aF(s)\} = \mathcal{L}^{-1}\{a\mathcal{L}\{f(t)\}\} = \mathcal{L}^{-1}\{\mathcal{L}\{af(t)\}\} = af(t) \tag{i}$$

We also have the relation

$$\mathcal{L}^{-1}\{\mathcal{L}\{f(t)\}\} = f(t) \tag{ii}$$

The ratio of Eqs. (i) and (ii) gives

$$\frac{\mathcal{L}^{-1}\{\mathcal{L}\{af(t)\}\}}{\mathcal{L}^{-1}\{\mathcal{L}\{f(t)\}\}} = a \tag{iii}$$

so that

$$\mathcal{L}^{-1}\{\mathcal{L}\{af(t)\}\} = a\mathcal{L}^{-1}\{\mathcal{L}\{f(t)\}\} \tag{iv}$$

Using $\mathcal{L}\{af(t)\} = a\mathcal{L}\{f(t)\}$ and $F(s) = \mathcal{L}\{f(t)\}$ in Eq. (iv) gives

$$\mathcal{L}^{-1}\{\mathcal{L}\{af(t)\}\} = \mathcal{L}^{-1}\{a\mathcal{L}\{f(t)\}\} = \mathcal{L}^{-1}\{aF(s)\} = a\mathcal{L}^{-1}\{F(s)\} \tag{v}$$

which is the desired result.

EXERCISE 9.10: What is the Laplace transform of the complex function $e^{i\omega t}$ with variable t and constant ω? Use the Laplace transform of $e^{i\omega t}$ and Euler's equation $e^{i\omega t} = \cos \omega t + i \sin \omega t$ to verify the Laplace transforms $\mathcal{L}\{\cos \omega t\} = s/(s^2 + \omega^2)$ and $\mathcal{L}\{\sin \omega t\} = \omega/(s^2 + \omega^2)$.

Solution: The Laplace transform of $e^{i\omega t}$ is

$$\mathcal{L}\{e^{i\omega t}\} = \int_0^\infty e^{-st} e^{i\omega t}\, dt = \int_0^\infty e^{-(s-i\omega)t}\, dt \qquad (i)$$

Evaluating the integral gives

$$\int_0^\infty e^{-(s-i\omega)t}\, dt = -\frac{1}{s-i\omega} e^{-(s-i\omega)t}\Big|_0^\infty = -\frac{1}{s-i\omega}(0-1) = \frac{1}{s-i\omega} \qquad (ii)$$

Equation (ii) can be rewritten in the form

$$\int_0^\infty e^{-(s-i\omega)t}\, dt = \frac{1}{s-i\omega}\frac{s+i\omega}{s+i\omega} = \frac{s+i\omega}{s^2+\omega^2} \qquad (iii)$$

Combining Eqs. (i) and (iii) gives the Laplace transform of $e^{i\omega t}$:

$$\mathcal{L}\{e^{i\omega t}\} = \int_0^\infty e^{-st} e^{i\omega t}\, dt = \frac{s+i\omega}{s^2+\omega^2} \qquad (iv)$$

If we now use Euler's equation in Eq. (iv) we find

$$\mathcal{L}\{e^{i\omega t}\} = \mathcal{L}\{\cos\omega t + i\sin\omega t\} = \mathcal{L}\{\cos\omega t\} + i\mathcal{L}\{\sin\omega t\} = \frac{s+i\omega}{s^2+\omega^2} \qquad (v)$$

The real part of Eq. (v) gives

$$\mathrm{Re}(\mathcal{L}\{e^{i\omega t}\}) = \mathcal{L}\{\cos\omega t\} = \frac{s}{s^2+\omega^2} \qquad (vi)$$

and the imaginary part gives

$$\mathrm{Im}(\mathcal{L}\{e^{i\omega t}\}) = \mathcal{L}\{\sin\omega t\} = \frac{\omega}{s^2+\omega^2} \qquad (vii)$$

Equations (vi) and (vii) verify the Laplace transforms $\mathcal{L}\{\cos\omega t\} = s/(s^2+\omega^2)$ and $\mathcal{L}\{\sin\omega t\} = \omega/(s^2+\omega^2)$.

S10   ORDINARY DIFFERENTIAL EQUATIONS

EXERCISE 10.1: Solve

$$\frac{dy}{dx} - y = e^{2x}$$

Solution: In this case, we have $f(x) = -1$ and $r(x) = e^{2x}$. Using $f(x)$ in h gives

$$h = \int f(x)\,dx = -x$$

Substituting this result into the equation for the general solution gives

$$y(x) = e^x\left[\int e^{-x}e^{2x}dx + C\right] = e^x[e^x + C] = Ce^x + e^{2x}$$

where C is a constant of integration. To verify the solution, we calculate

$$\frac{dy}{dx} = Ce^x + 2e^{2x}$$

and find the expected result

$$\frac{dy}{dx} - y = Ce^x + 2e^{2x} - Ce^x - e^{2x} = e^{2x}$$

EXERCISE 10.2: Solve

$$\frac{dy}{dx} + y\tan x = \sin 2x, \quad y(0) = 1$$

Solution: We first observe that

$$f(x) = \tan x, \quad r(x)\sin 2x$$

so that

$$h = \int f(x)\,dx = \int \tan x\,dx = \ln(\sec x)$$

Using this result gives

$$e^h = \exp[\ln(\sec x)] = \sec x$$
$$e^{-h} = (e^h)^{-1} = (\sec x)^{-1} = \cos x$$

and

$$e^h r = \sec x \sin 2x = \frac{1}{\cos x}(2\sin x\cos x) = 2\sin x$$

The general solution becomes

$$y(x)\cos x\left[2\int \sin x\,dx + C\right] = C\cos x - 2\cos^2 x$$

The boundary condition $y(0) = 1$ implies $1 = C - 2$ or $C = 3$. Thus

$$y(x) = 3\cos x - 2\cos^2 x$$

The solution is verified by evaluating

$$\frac{dy}{dx} = 3(-\sin x) - 4\cos x(-\sin x) = -3\sin x + 4\cos x \sin x$$

Using the trigonometric identity

$$\sin 2x = 2\sin x \cos x$$

we get

$$\frac{dy}{dx} = -3\sin x + 2\sin 2x$$

Now calculate the right-hand side to get the expected result

$$\frac{dy}{dx} + y\tan x = -3\sin x + 2\sin 2x + (3\cos x - 2\cos^2 x)\frac{\sin x}{\cos x}$$

$$= -3\sin x + 2\sin 2x + 3\sin x - 2\cos x \sin x$$

$$= 2\sin 2x - \sin 2x$$

$$= \sin 2x$$

EXERCISE 10.3: Convert the nonautonomous equation

$$\frac{dx_1}{dt} = 1 - t + 4x_1, \; x_1(0) = 1$$

to an autonomous system.

Solution: The single-equation problem with an initial condition is equivalent to the two-equation problem

$$\frac{dx_1}{dt} = 1 - x_2 + 4x_1$$

$$\frac{dx_2}{dt} = 1$$

subject to

$$x_1(0) = 1, \; x_2(0) = 0$$

EXERCISE 10.4: Convert the FORTRAN 90/95 program for the Runge–Kutta algorithm in the above example to the C++ programming language. Show that the numerical and analytical results agree.

Solution: The following code illustrates the Runge–Kutta algorithm in C++. It may be necessary to modify the code to work with different C++ compilers. It reproduces both the numerical and analytical results obtained by the FORTRAN 90/95 program in Chapter 10.

RUNGE–KUTTA ALGORITHM IN C++

SOURCE: A. C. FANCHI

```cpp
/*
 * 4th-Order Runge-Kutta Algorithm in C++
 */

#include <math.h>
#include <stdio.h>

float F1 (float, float, float);
float F2 (float, float, float);

int main (int argc, char* argv[])
{
//Initial Conditions
const float T0=0, X10=0, X20=1, H=0.1;
float X1, X2, T, X1AN, X2AN;
float A1, A2, B1, B2, C1, C2, D1, D2;
int NS=20;
int counter=1;
//Initialize Variables
X1=X10;
X2=X20;
T=T0;

//Create and open output file
FILE* outfile=fopen("RK_4th_C++.out", "w");

//Output file header
fprintf(outfile, "NUMERICAL ANALYTICAL\n");
fprintf(outfile, "STEP T X1 X2 X1AN X2AN\n");
```

```
do
{
A1=F1 (X1,X2,T);
A2=F2 (X1, X2, T);
B1=F1 (X1+(H*A1/2),X2+(H*A2/2), T+(H/2));
B2=F2 (X1+(H*A1/2),X2+(H*A2/2), T+(H/2));
C1=F1 (X1+(H*B1/2),X2+(H*B2/2), T+(H/2));
C2=F2 (X1+(H*B1/2),X2+(H*B2/2), T+(H/2));
D1=F1 (X1+(H*C1),X2+(H*C2), T+(H));
D2=F2 (X1+(H*C1),X2+(H*C2), T+(H));
//Update X1, X2, T
X1=X1+(H/6) * (A1+2*B1+2*C1+D1);
X2=X2+(H/6) * (A2+2*B2+2*C2+D2);
T=T+H;
X1AN=sin(T);
X2AN=cos(T);
//Output data to file
fprintf (outfile, "%4d %9f %9f %9f %9f %9f\n", counter,
T, X1, X2, X1AN, X2AN);
counter++;
} while (counter <= NS);

//Close output file
fclose(outfile);

return 0;
}


float F1 (float X1, float X2, float T)
{
float F1;
F1=X2;
return F1;
}


float F2 (float X1, float X2, float T)
{
float F2;
F2=-X1;
return F2;
}
```

EXERCISE 10.5: Write the third-order ODE

$$\frac{d^3y}{dt^3} = 2y\left(\frac{dy}{dt}\right)^2 + 4y$$

as a first-order system.

Solution: In this case $n = 3$ for the third-order ODE. Define the variables

$$x_1 = y, x_2 = \frac{dy}{dt}, x_3 = \frac{d^2y}{dt^2}$$

The resulting system of equations is

$$\frac{dx_1}{dt} = x_2$$

$$\frac{dx_2}{dt} = x_3$$

$$\frac{dx_3}{dt} = 2x_1 \, x_2^2 + 4x_1$$

EXERCISE 10.6: The linear harmonic oscillator is governed by

$$\frac{d^2y}{dt^2} + y = 0, \ y(0) = a, \ \left.\frac{dy}{dt}\right|_0 = b$$

Convert this second-order ODE to a first-order system.

Solution: In this case $n = 2$ for the second-order ODE. Define the variables

$$x_1 = y, x_2 = \frac{dy}{dt}$$

so that

$$\frac{dx_1}{dt} = x_2$$

$$\frac{dx_2}{dt} = -x_1$$

and the initial conditions are represented as

$$x_1(0) = a, \ x_2(0) = b$$

EXERCISE 10.7: Suppose we are given the equation $m\ddot{x} + c\dot{x} + kx = 0$ as a model of a mechanical system, where $\{m, c, k\}$ are real constants. Under what conditions are the solutions stable? *Hint*: Rearrange the equation to look like Eq. (10.3.1) and then find the eigenvalues corresponding to the characteristic equation of the stability analysis.

Solution: Rewrite the ODE as

$$\ddot{x} + \frac{c}{m}\dot{x} + \frac{k}{m}x = 0$$

so that $\gamma_1 = c/m$ and $\gamma_0 = k/m$ by comparison with Eq. (10.3.1) of the stability analysis presentation. The displacement of the solution from equilibrium is

$$\boldsymbol{u} = e^{\lambda_{\pm}t}\boldsymbol{g}$$

where the eigenvalues are given by

$$\lambda_{\pm} \frac{1}{2}\left[-\gamma_1 \pm \sqrt{\gamma_1^2 - 4\gamma_0}\right] = \frac{1}{2}\left[-\frac{c}{m} \pm \sqrt{\frac{c^2}{m^2} - \frac{4k}{m}}\right]$$

The displacement converges as long as $\lambda_{\pm} < 0$; thus the physical constants must satisfy the criterion

$$-\frac{c}{m} \pm \sqrt{\frac{c^2}{m^2} - \frac{4k}{m}} < 0$$

for stability. Notice that the stability criterion is not satisfied if $c = 0$. In this case we have a wave equation and the solution of the ODE is undamped. A detailed solution of this ODE is presented in Chapter 11. If $k = 0$, the stability criterion is

$$-\frac{c}{m} \pm \frac{c}{m} < 0$$

All the solutions are stable for $\lambda_- = c/m$ and m positive and $k = 0$.

EXERCISE 10.8: Another way to see the onset of chaos is to plot the number of fixed points versus the parameter λ. This plot is called the *logistic map*. Plot the logistic map for the May equation in the range $3.5 \le \lambda \le 4.0$ and initial condition $x_0 = 0.5$.

Solution: The logistic map for the May equation is shown in the accompanying figure. Fifty iterations were made for each value of λ. Values of λ were increased by an increment of 0.01.

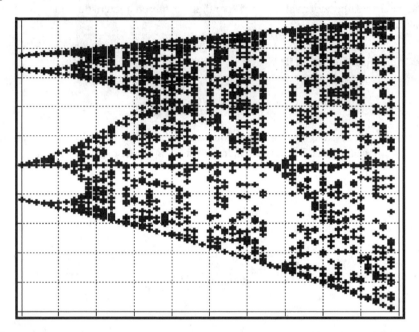

S11 ODE SOLUTION TECHNIQUES

EXERCISE 11.1: Application to an Electrical System—*RLC* Circuit. Find a particular solution of the nonhomogeneous ODE

$$L\ddot{I} + R\dot{I} + \frac{I}{\hat{C}} = E_0\omega\cos\omega t$$

where L = inductance of inductor, R = resistance of resistor, $\hat{C}$ = capacitance of capacitor, $E_0\omega$ = maximum value of derivative of electromotive force (E_0 = voltage at $t = 0$), and $I(t)$ = current.

Solution: The *RLC* equation has the form

$$A\ddot{x} + B\dot{x} + Cx = D\cos\omega t$$

where $A = L$, $B = R$, $C = 1/\hat{C}$, $D = E_0\omega$, and $x = I(t)$. The particular solution is

$$I(t) = a\cos\omega t + b\sin\omega t$$

with

$$a = E_0\omega \frac{\left(\frac{1}{\hat{C}} - L\omega^2\right)}{\left(\frac{1}{\hat{C}} - L\omega^2\right)^2 + R^2\omega^2}$$

and

$$b = E_0\omega \frac{R\omega}{\left(\frac{1}{\hat{C}} - L\omega^2\right)^2 + R^2\omega^2}$$

Factor $1/\omega^2$ to get

$$a = \frac{E_0\left(\frac{1}{\omega\hat{C}} - L\omega\right)}{\left(\frac{1}{\omega\hat{C}} - L\omega\right)^2 + R^2}$$

and

$$b = \frac{E_0 R}{\left(\frac{1}{\omega\hat{C}} - L\omega\right)^2 + R^2}$$

These expressions are transformed into a physically more interesting form by defining reactance $S = L\omega - (1/\omega\hat{C})$ to obtain

$$a = -\frac{E_0 S}{S^2 + R^2}$$

$$b = \frac{E_0 R}{S^2 + R^2}$$

Impedance is defined as $\sqrt{R^2 + S^2}$ and corresponds to resistance for an AC system. See Exercise 2.2 of Chapter 2 for more details.

EXERCISE 11.2: Find the general solution of the one-dimensional Schrödinger equation

$$-\frac{\hbar^2}{2m}\frac{d^2}{dx^2}\psi(x) - E\psi(x) = V(x)$$

for the wavefunction $\psi(x)$. The two boundary conditions are $\psi(x_0) = 0$, $\psi'(x_0) = 0$.

Solution: The one-dimensional Schrödinger equation may be put in a more familiar mathematical form by rearranging it and writing it as

$$\frac{d^2\psi(x)}{dx^2} + a_0\psi(x) = h(x)$$

where

$$a_0 = \frac{2mE}{\hbar^2}, \, h(x) = -\frac{2m}{\hbar^2}V(x), \, a_1(x) = 0$$

The homogeneous equation is

$$\frac{d^2\psi_h(x)}{dx^2} = -a_0\psi_h(x)$$

Using the trial solution

$$\psi_h(x) = c_1 \sin kx + c_2 \cos kx$$

in the homogeneous equation gives

$$\frac{d^2\psi_h(x)}{dx^2} = -k^2[c_1 \sin kx + c_2 \cos kx] = -a_0\psi_h(x)$$

so that

$$k^2 = a_0$$

or

$$k = \sqrt{2mE/\hbar^2}$$

To solve the nonhomogeneous equation, we calculate the Green's function

$$K(x,t) = \frac{\sin kt \cos kx - \sin kx \cos kt}{\sin kt(-\sin kt) - \cos kt(\cos kt)}$$

$$= \frac{\sin kt \cos kx - \sin kx \cos kt}{-[\sin^2 kt + \cos^2 kt]}$$

or

$$K(x,t) = \sin kx \cos kt - \sin kt \cos kx$$

where t is the dummy variable of integration. The particular solution of the nonhomogeneous equation is

$$\psi_p(x) = \int_{x_0}^{x} K(x,t)\, h(t)\, dt$$

$$= \int_{x_0}^{x} [\sin kx \cos kt - \sin kt \cos kx] \left(-\frac{2\, mV(t)}{\hbar^2} \right) dt$$

$$= \int_{x_0}^{x} \sin(kx - kt) \left(-\frac{2\, mV(t)}{\hbar^2} \right) dt$$

The general solution is the sum of the homogeneous and particular solutions, that is,

$$\psi(x) = \psi_p(x) + \psi_h(x)$$

An explicit solution may be obtained once $V(x)$ is specified.

EXERCISE 11.3: Use the Frobenius method to solve the equation

$$\left[\frac{d^2}{dr^2} + \frac{2}{r}\frac{d}{dr} - \frac{\varepsilon}{r^2} + \omega^2 \right] R(r) = 0 \tag{i}$$

for $R(r)$ when ε and ω are independent of the variable r, and $r > 0$.

Solution: Multiply Eq. (i) by r^2 to find

$$r^2 R'' + 2rR' + (\omega^2 r^2 - \varepsilon)R = 0 \tag{ii}$$

where the primes denote differentiation with respect to r. Assume a Frobenius series solution of the form

$$R(r) = \sum_{k=0}^{\infty} a_k\, r^{k+\lambda} \tag{iii}$$

where the constant λ is now undetermined. The first and second derivatives are given by

$$R'(r) = \sum_{k=0}^{\infty} a_k (k + \lambda) r^{k+\lambda-1} \tag{iv}$$

and

$$R''(r) = \sum_{k=0}^{\infty} a_k (k + \lambda)(k + \lambda - 1) r^{k+\lambda-2} \tag{v}$$

Equation (ii) becomes

$$\sum_{k=0}^{\infty} a_k(k+\lambda)(k+\lambda-1)r^{k+\lambda} + 2\sum_{k=0}^{\infty} a_k(k+\lambda)r^{k+\lambda}$$
$$+ \sum_{k=0}^{\infty} \omega^2 a_k r^{k+\lambda+2} - \varepsilon \sum_{k=0}^{\infty} a_k r^{k+\lambda} = 0 \qquad \text{(vi)}$$

or

$$\sum_{k=0}^{\infty} a_k[(k+\lambda)(k+\lambda-1) + 2(k+\lambda) - \varepsilon]r^{k+\lambda}$$
$$+ \sum_{k=0}^{\infty} a_k \omega^2 r^{k+\lambda+2} = 0 \qquad \text{(vii)}$$

Expanding the first few terms of Eq. (vii) gives

$$a_0[\lambda(\lambda-1) + 2\lambda - \varepsilon]r^{\lambda}$$
$$+a_1[(1+\lambda)(\lambda) + 2(1+\lambda) - \varepsilon]r^{\lambda+1}$$
$$+\{a_2[(2+\lambda)(\lambda+1) + 2(2+\lambda) - \varepsilon] + a_0\omega^2\}r^{\lambda+2}$$
$$+\{a_3[(3+\lambda)(\lambda+2) + 2(3+\lambda) - \varepsilon] + a_1\omega^2\}r^{\lambda+3} + \cdots = 0 \qquad \text{(viii)}$$

The indicial equation is found by setting the coefficient of r^{λ} to 0 in the $k=0$ term; thus

$$a_0[\lambda(\lambda-1) + 2\lambda - \varepsilon] = 0 \qquad \text{(ix)}$$

A nontrivial solution of Eq. (ix); that is, a solution in which $a_0 \neq 0$, gives the roots

$$\lambda_{\pm} = \frac{-1 \pm \sqrt{1+4\varepsilon}}{2} \qquad \text{(x)}$$

The coefficient of the $r^{\lambda+1}$ term in Eq. (viii) must satisfy

$$a_1[\lambda^2 + 3\lambda + 2 - \varepsilon] = 0 \qquad \text{(xi)}$$

Since λ was determined in Eq. (x), the only way to satisfy Eq. (xi) is to require $a_1 = 0$.

The recurrence relation for the nonzero expansion coefficients is found from Eq. (viii) to be

$$a_{k+2} = -\frac{\omega^2}{\beta_\pm(k)} a_k \tag{xii}$$

where

$$\beta_\pm(k) = (k+2+\lambda_\pm)(k+1+\lambda_\pm) + 2(k+2+\lambda_\pm) - \varepsilon \tag{xiii}$$

and $k = 2,4,6,\ldots$.

EXERCISE 11.4: Use the Laplace transform to solve the ODE

$$y'' + (a+b)y' + aby = 0 \tag{i}$$

where the prime denotes differentiation with respect to x, and the boundary conditions are

$$y(0) = \alpha, \; y'(0) = \beta \tag{ii}$$

The quantities a, b, α, β are constants.

Solution: The Laplace transform of Eq. (i) is

$$s^2 \mathcal{L}(s) - sy(0) - y'(0) + (a+b)[s\mathcal{L}(s) - y(0)] + ab\mathcal{L}(s) = 0 \tag{iii}$$

where $\mathcal{L}(s) = \mathcal{L}\{y(x)\}$. Substituting Eq. (ii) into Eq. (iii) gives

$$s^2 \mathcal{L}(s) - s\alpha - \beta + (a+b)s\mathcal{L}(s) - \alpha(a+b) + ab\mathcal{L}(s) = 0$$

Rearranging lets us write

$$\left[s^2 + (a+b)s + ab\right]\mathcal{L}(s) = \alpha s + \alpha(a+b) + \beta$$

or

$$\mathcal{L}(s) = \frac{\alpha s + \alpha(a+b) + \beta}{s^2 + (a+b)s + ab}$$

$$= \frac{\alpha s + \alpha(a+b) + \beta}{(s+a)(s+b)} \tag{iv}$$

Assume $\mathcal{L}(s)$ can be written in the form

$$\mathcal{L}(s) = \frac{p}{s+a} - \frac{q}{s+b} \tag{v}$$

Equating Eqs. (iv) and (v) gives

$$q = \frac{a\alpha + \beta}{b - a}$$

$$p = \alpha + q = \frac{\alpha b + \beta}{b - a} \qquad \text{(vi)}$$

Inverting $\mathcal{L}(s)$ yields the solution

$$y(x) = \mathcal{L}^{-1}\left\{\frac{p}{s + a}\right\} - \mathcal{L}^{-1}\left\{\frac{q}{s + b}\right\}$$

or

$$y(x) = pe^{-ax} - qe^{-bx} \qquad \text{(vii)}$$

Replacing p and q by the expressions in Eq. (vi) gives

$$y(x) = \frac{\alpha b + \beta}{b - a} e^{-ax} - \frac{a\alpha + \beta}{b - a} e^{-bx} \qquad \text{(viii)}$$

Equation (viii) is validated by substitution into Eq. (i) and application of Eq. (ii).

S12 PARTIAL DIFFERENTIAL EQUATIONS

EXERCISE 12.1: Assume the boundary conditions of Laplace's equation are

$$\Phi(0, y, z) = \Phi(x_B, y, z) = 0$$
$$\Phi(x, 0, z) = \Phi(x, y_B, z) = 0$$
$$\Phi(x, y, 0) = 0$$
$$\Phi(x, y, z_B) = \Phi_B(x, y)$$

in the intervals $0 \le x \le x_B$, $0 \le y \le y_B$ and $0 \le z \le z_B$. Find the solution of Laplace's equation using these boundary conditions and the trigonometric solutions in Eq. (12.3.7).

Solution: The boundary condition at $x = 0$ implies

$$X(0) = 0 = a_2'$$

The boundary condition at $x = x_B$ together with the above result implies

$$X(x_B) = 0 = a_1' \sin \alpha x_B$$

This relationship is satisfied for a nonzero value of a_1' when $\alpha x_B = n\pi$ and $n = 1, 2, 3, \ldots$. The integers $0, -1, -2, \ldots$ also satisfy the condition at $X(x_B)$. The choice of positive integers has been made in anticipation of using a Fourier series expansion later in the solution. Similar results to the x coordinate solution $X(x)$ are obtained for the y coordinate solution $Y(y)$ by imposing the y coordinate boundary conditions. The results are

$$X(x) = a_1' \sin(\alpha_n x), \quad \alpha_n = \frac{n\pi}{x_B}$$

$$Y(y) = b_1' \sin(\beta_m y), \quad \beta_m = \frac{m\pi}{y_B}$$

where n and m are positive integers. Part of the z coordinate solution $Z(z)$ is found by applying the boundary condition at $z = 0$ to get

$$Z(z) = c_1' \sinh(\delta_{nm} z), \quad \delta_{nm}^2 = \alpha_n^2 + \beta_m^2$$

A partial solution of the Laplace equation is the product XYZ, or

$$\Phi_{nm}(x, y, z) = V_{nm} \sin\left(\frac{n\pi}{x_B}\right) \sin\left(\frac{m\pi}{y_B}\right) \sinh(\delta_{nm} z)$$

where the product of constants $a_1' b_1' c_1'$ has been replaced by the constant V_{nm}. The general solution of the Laplace equation is a superposition, or sum, of partial solutions

$$\Phi(x, y, z) = \sum_{n=1}^{\infty} \sum_{m=1}^{\infty} \Phi_{nm}(x, y, z)$$

The last boundary condition at $z = z_B$ is satisfied by finding expansion coefficients $\{V_{nm}\}$, such that

$$\Phi_B(x, y) = \sum_{n=1}^{\infty} \sum_{m=1}^{\infty} \Phi_{nm}(x, y, z_B)$$

$$= \sum_{n=1}^{\infty} \sum_{m=1}^{\infty} V_{nm} \sin(\alpha_n x) \sin(\beta_m y) \sinh(\delta_{nm} z_B)$$

This is a Fourier series representation of $\Phi_B(x, y)$, and the coefficients $\{V_{nm}\}$ are given by [Jackson, 1962]

$$V_{nm} = \frac{4}{x_B y_B \sinh(\delta_{nm} z_B)} \int_0^{x_B} \int_0^{y_B} \Phi_B(x, y) \sin(\alpha_n x) \sin(\beta_m y) \, dx \, dy$$

EXERCISE 12.2: Separation of Variables. Find the solution of

$$\frac{\partial^2 \psi}{\partial x^2} = \frac{\partial \psi}{\partial t} \tag{i}$$

by assuming

$$\psi(x, t) = X(x) T(t) \tag{ii}$$

Solution: Substituting Eq. (ii) into Eq. (i) gives

$$T(t) \frac{\partial^2 X(x)}{\partial x^2} = X(x) \frac{\partial T(t)}{\partial t}$$

Divide by XT to get

$$\frac{1}{X} \frac{\partial^2 X}{\partial x^2} = \frac{1}{T} \frac{\partial T}{\partial t} = \kappa^2$$

where κ^2 is the separation constant. Note that each side of the above equation must be constant if the equation is satisfied for arbitrary values of the variables x, t. The resulting ODEs are

$$\frac{d^2 X}{dx^2} = \kappa^2 X \tag{iii}$$

$$\frac{dT}{dt} = \kappa^2 T \tag{iv}$$

Equation (iii) has the solution

$$X(x) = a e^{\kappa x} + b e^{-\kappa x}$$

and Eq. (iv) has the solution

$$T(t) = c e^{\kappa^2 t}$$

Combining these solutions with Eq. (ii) gives

$$\psi(x, t) = (a e^{\kappa x} + b e^{-\kappa x}) c e^{\kappa^2 t}$$

or

$$\psi(x, t) = a' e^{\kappa x + \kappa^2 t} + b' e^{-\kappa x + \kappa^2 t}$$

where a', b' are undetermined constants. The constants a', b', κ are determined by imposing two boundary conditions and one initial condition.

EXERCISE 12.3: Find $u(kx - \omega t)$ using Eq. (12.3.13) and assuming $n = 1$. The nonlinear equation in this case is known as *Burger's equation*.

Solution: Equation (12.3.13) with $n = 1$ is

$$z = \int \frac{dg}{c + \left(\frac{ak - \omega}{Dk^2}\right)g + \frac{b}{2Dk}g^2} = \int \frac{dg}{\alpha + \beta g + \gamma g^2}$$

where β and γ are the coefficients of g and g^2, respectively, and $\alpha = c$. The solution of the integral depends on the values of α, β, γ and the parameter $q = 4\alpha\gamma - \beta^2$.
    If $q > 0$, then

$$z = \int \frac{dg}{\alpha + \beta g + \gamma g^2} = \frac{2}{\sqrt{q}} \tan^{-1}\left(\frac{2\gamma g + \beta}{\sqrt{q}}\right)$$

or

$$g(z) = \frac{1}{2\gamma}\left[\sqrt{q}\tanh\left(\frac{\sqrt{q}\,z}{2}\right) - \beta\right]$$

Recall that $u(kx - \omega t) = g(x)$ with the variable transformation $z = kx - \omega t$.
    If $q < 0$, then

$$z = \int \frac{dg}{\alpha + \beta g + \gamma g^2} = -\frac{2}{\sqrt{-q}} \tanh^{-1}\left(\frac{2\gamma g + \beta}{\sqrt{-q}}\right)$$

or

$$g(z) = \frac{1}{2\gamma}\left[\sqrt{-q}\,\tanh\left(\frac{\sqrt{-q}\,z}{-2}\right) - \beta\right]$$

If $q = 0$, then

$$z = \int \frac{dg}{\left(\frac{\beta}{2\gamma} + g\right)^2} = \frac{1}{\gamma}\left[-\frac{1}{\left(\frac{\beta}{2\gamma} + g\right)}\right] = -\frac{1}{\frac{\beta}{2} + \gamma g}$$

or

$$g(z) = -\frac{1}{\gamma}\left[\frac{\beta}{2} + \frac{1}{z}\right]$$

EXERCISE 12.4: Find the dispersion relation associated with the Klein–Gordon equation

$$m^2 \phi(x, t) = -\left[\frac{\partial^2}{\partial t^2} - \frac{\partial^2}{\partial x^2}\right]\phi(x, t)$$

Hint: Use a solution of the form $\phi(x, t) = \iint \eta(k, \omega)\, e^{i(\omega t - kx)} dk\, d\omega$.

Solution: Substitute $\phi(x, t)$ into the Klein–Gordon equation and calculate the derivatives to get

$$\iint \eta(k, \omega)(m^2)e^{i(\omega t - kx)} dk\, d\omega$$

$$= \iint \eta(k, \omega)\{-[-\omega^2 - (-k^2)]\}e^{i(\omega t - kx)} dk\, d\omega$$

Combining like terms in this equation lets us write

$$\iint \eta(k, \omega)\{\lfloor m^2 - \omega^2 + k^2 \rfloor\}e^{i(\omega t - kx)} dk\, d\omega = 0$$

which is satisfied by the dispersion relation $\omega^2 = m^2 + k^2$. In this case, ω and k are variables and m is a constant.

S13 INTEGRAL EQUATIONS

EXERCISE 13.1: Write the linear oscillator equation $y'' + \omega^2 y = 0$ with initial conditions $y(0) = 0$, $y'(0) = 1$ in the form of an integral equation.

Solution: The initial conditions at the point $x = a = 0$ are $y(0) = y_0 = 0$, $y'(0) = y_0' = 1$. The kernel is

$$K(x, t) = (t - x)[B(t) - A'(t)] - A(t) = (t - x)\omega^2$$

from Eq. (13.2.13), and the function $f(x)$ is

$$f(x) = y_0'(x - a) + y_0 = x$$

from the initial conditions and Eq. (13.2.14). The formal solution of the linear oscillator equation $y'' + \omega^2 y = 0$ is

$$y(x) = f(x) + \int_a^x K(x, t)\, y(t)\, dt = x + \omega^2 \int_0^x (t - x)\, y(t)\, dt$$

EXERCISE 13.2: Solve the linear oscillator equation $y'' + \omega^2 y = 0$ with initial conditions $y(0) = 0$, $y(b) = 0$. *Hint*: Modify the above procedure to account for the use of boundary conditions $y(0) = 0$, $y(b) = 0$ [see Arfken and Weber, 2001, pp. 987–989].

Solution: Integrate $y'' + \omega^2 y = 0$ with respect to x between the lower and upper limits 0 and x, and respectively:

$$y' - y_0'(0) = -\omega^2 \int_0^x y \, dx \tag{i}$$

Integrate again:

$$y - y(0) - y'(0)x = -\omega^2 \int_0^x \int_0^x y(t)dt \, dx \tag{ii}$$

Use Eq. (13.2.8) to write

$$\int_0^x \left\{ \int_0^x y(t)dt \right\} dx = \int_0^x \left\{ \int_0^x y(t)dx \right\} dt = \int_0^x (x - t)f(t)dt \tag{iii}$$

Using Eq. (iii) in Eq. (ii) gives

$$y(x) = y(0) + y'(0)x - \omega^2 \int_0^x (x - t) y(t)dt \tag{iv}$$

Substituting the initial conditions $y(0) = 0$, $y(b) = 0$ into Eq. (iv) lets us write

$$y(b) = 0 = y(0) + y'(0)b - \omega^2 \int_0^x (b - t) y(t)dt$$

$$= y'(0)b - \omega^2 \int_0^x (b - t) y(t)dt \tag{v}$$

We solve Eq. (v) for $y'(0)$; thus

$$y'(0) = \frac{\omega^2}{b} \int_0^x (b - t) y(t)dt \tag{vi}$$

Substituting Eq. (vi) into Eq. (iv) yields

$$y(x) = \frac{\omega^2 x}{b} \int_0^b (b - t) y(t)dt - \omega^2 \int_0^x (x - t) y(t)dt \tag{vii}$$

Break the interval $[0, b]$ into the two intervals $[0, x]$ and $[x, b]$:

$$y(x) = \frac{\omega^2 x}{b}\left[\int_0^x (b-t)\,y(t)dt + \int_x^b (b-t)\,y(t)dt\right] - \omega^2 \int_0^x (x-t)\,y(t)dt \qquad \text{(viii)}$$

Combine the first and third integrals on the right-hand side of Eq. (viii) and rearrange:

$$y(x) = \frac{\omega^2 x}{b}\int_x^b (b-t)\,y(t)\,dt + \omega^2 \int_0^x \left[\frac{x}{b}(b-t)-(x-t)\right]y(t)\,dt \qquad \text{(ix)}$$

Use

$$\frac{x}{b}(b-t)-(x-t) = x - x\frac{t}{b} - x + t = \frac{t}{b}(b-x)$$

to simplify Eq. (ix):

$$y(x) = \omega^2 \int_x^b \frac{x}{b}(b-t)\,y(t)dt + \omega^2 \int_0^x \left[\frac{t}{b}(b-x)\right]y(t)dt \qquad \text{(x)}$$

The kernel in Eq. (x) can be written as

$$K(x, t) = \begin{cases} \dfrac{t}{b}(b-x), \ t < x \\[2mm] \dfrac{x}{b}(b-t), \ x < t \end{cases} \qquad \text{(xi)}$$

so that the solution $y(x)$ has the form

$$y(x) = \omega^2 \int_0^b K(x, t)\,y(t)dt \qquad \text{(xii)}$$

Notice that the kernel is symmetric, $K(x, t) = K(t, x)$, and continuous:

$$\left.\frac{t}{b}(b-x)\right|_{t=x} = \left.\frac{x}{b}(b-t)\right|_{x=t}$$

The kernel $K(x, t)$ is a Green's function. The derivative $\partial K(x, t)/\partial t$ of the kernel with respect to t is discontinuous at $t = x$.

EXERCISE 13.3: Solve the homogeneous Fredholm equation $\phi(x) = \lambda \int_{-1}^{1} (t+x)\,\phi(t)\,dt$ [see Arfken and Weber, 2001, p. 1002].

Solution: Since the equation is homogeneous, we have $f(x) = 0$ in Eq. (13.3.1) so the resulting equation is $\phi(x) = \lambda \int_a^b K(x, t)\phi(t)\,dt$. We construct a separable kernel by writing

$$M_1 = 1, M_2(x) = x, N_1(t) = t, N_2 = 1$$

in Eq. (13.4.1) so that

$$K(x, t) = \sum_{j=1}^{2} M_j(x)N_j(t) = M_1 N_1 + M_2 N_2 = t + x$$

The coefficients $b_i = \int_a^b N_i(x)f(x)\,dx$ are $b_1 = 0$, $b_2 = 0$ since $f(x) = 0$. The set of coefficients $a_{ij} = \int_a^b N_i(x)M_j(x)\,dx$ are given by

$$a_{11} = a_{22} = \int_{-1}^{1} x\,dx = \left.\frac{x^2}{2}\right|_{-1}^{1} = 0$$

$$a_{12} = \int_{-1}^{1} x^2\,dx = \left.\frac{x^3}{3}\right|_{-1}^{1} = \frac{2}{3}$$

$$a_{21} = \int_{-1}^{1} dx = x|_{-1}^{1} = 2$$

The determinant $|\boldsymbol{I} - \lambda \boldsymbol{A}| = 0$ for the eigenvalues is

$$\begin{vmatrix} 1 - \lambda \cdot 0 & -\frac{2}{3}\lambda \\ -2\lambda & 1 - \lambda \cdot 0 \end{vmatrix} = 1 - \frac{4}{3}\lambda^2 = 0$$

The resulting eigenvalues are

$$\lambda_{\pm} = \pm\sqrt{\frac{3}{4}}$$

Expanding the matrix equation $(\boldsymbol{I} - \lambda \boldsymbol{A})\boldsymbol{c} = \boldsymbol{b}$ gives

$$(1 - \lambda a_{11})c_1 + (-\lambda a_{12})c_2 = 0$$
$$(-\lambda a_{21})c_1 + (1 - \lambda a_{22})c_2 = 0$$

Solving for c_1 in terms of c_2 provides the following two relations:

$$c_1 - c_2\lambda\frac{2}{3} = 0 \Rightarrow c_1 = \frac{2}{3}\lambda c_2$$

$$-\lambda 2 c_1 + c_2 = 0 \Rightarrow c_1 = \frac{c_2}{2\lambda}$$

Substituting in values of the eigenvalues gives

$$\lambda_+: c_1 - c_2\left(\sqrt{\frac{3}{4}}\right)\frac{2}{3} = 0 \Rightarrow c_1 = \frac{c_2}{\sqrt{3}}$$

$$\lambda_-: c_1 - c_2\left(-\sqrt{\frac{3}{4}}\right)\frac{2}{3} = 0 \Rightarrow c_1 = -\frac{c_2}{\sqrt{3}}$$

Combining relations gives

$$c_1 = \pm\frac{c_2}{\sqrt{3}}$$

Alternatively, solving for c_2 in terms of c_1 results in the relations

$$c_2 = \pm\sqrt{3}c_1$$

If we set $c_1 = 1$, then we find $c_2 = \pm\sqrt{3}$ and the solutions for both eigenvalues are determined using Eq. (13.4.3):

$$\lambda_1 = \lambda_+ = \frac{\sqrt{3}}{2}: \phi_1 = \lambda_1 \sum_{j=1}^{2} c_j M_j(x) = \frac{\sqrt{3}}{2}\left(1 + \sqrt{3}x\right)$$

$$\lambda_2 = \lambda_- = -\frac{\sqrt{3}}{2}: \phi_2 = \lambda_2 \sum_{j=1}^{2} c_j M_j(x) = -\frac{\sqrt{3}}{2}\left(1 - \sqrt{3}x\right)$$

EXERCISE 13.4: Use the Laplace transform to solve the integral equation

$$V_0 \cos \omega t = Ri(t) + \frac{1}{C}\int_0^t i(\tau)\,d\tau$$

for an RC circuit. An RC circuit consists of a resistor with resistance R and a capacitor with capacitance C in series with a time-varying voltage $V(t) = V_0 \cos \omega t$ that has frequency ω.

Solution: Rearrange the integral equation to the form

$$i(t) = \frac{V_0}{R}\cos \omega t - \frac{1}{RC}\int_0^t i(\tau)\,d\tau \tag{i}$$

Comparing Eq. (i) to Eq. (13.5.1) lets us write

$$\phi(t) = i(t)$$

$$f(t) = \frac{V_0}{R} \cos \omega t \tag{ii}$$

$$\lambda = -\frac{1}{RC}$$

To use Eq. (13.5.5), we evaluate the Laplace transform

$$\mathcal{L}\{f(t)\} = \frac{V_0}{R} \mathcal{L}\{\cos \omega t\} = \frac{V_0}{R} \frac{s}{s^2 + \omega^2} \tag{iii}$$

Substituting Eqs. (ii) and (iii) into Eq. (13.5.5) gives

$$i(t) = \mathcal{L}^{-1}\left\{\frac{s}{s-\lambda} \frac{V_0}{R} \frac{s}{s^2 + \omega^2}\right\} = \mathcal{L}^{-1}\left\{\frac{V_0}{R} \frac{s^2}{\left(s + \dfrac{1}{RC}\right)(s^2 + \omega^2)}\right\} \tag{iv}$$

The partial fraction expansion of the s-dependent term is

$$\frac{s^2}{\left(s + \dfrac{1}{RC}\right)(s^2 + \omega^2)} = \frac{a_1}{s + \dfrac{1}{RC}} + \frac{a_2 s + a_3}{s^2 + \omega^2}$$

$$= \frac{a_1(s^2 + \omega^2) + (a_2 s + a_3)\left(s + \dfrac{1}{RC}\right)}{\left(s + \dfrac{1}{RC}\right)(s^2 + \omega^2 t)} \tag{v}$$

where $\{a_1, a_2, a_3\}$ are expansion coefficients. Multiplying Eq. (v) by the denominator of the left-hand side lets us expand Eq. (v) in powers of s:

$$s^2 = a_1 s^2 + a_1 \omega^2 + a_2 s^2 + \frac{a_2 s}{RC} + a_3 s + \frac{a_3}{RC} \tag{vi}$$

Equating the coefficients of the powers of s gives

$$s^2: a_1 + a_2 = 1 \Rightarrow a_1 = 1 - a_2$$

$$s^1: \frac{a_2}{RC} + a_3 = 0 \Rightarrow a_2 = -a_3 RC \tag{vii}$$

$$s^0: a_1 \omega^2 + \frac{a_3}{RC} = 0 \Rightarrow a_3 = -a_1 RC \omega^2$$

Solving Eq. (vii) for $\{a_1, a_2, a_3\}$ lets us write

$$a_1 = 1/[1 + (RC\omega)^2]$$

$$a_2 = 1 - a_1 = [(RC\omega)^2]/[1 + (RC\omega)^2] \qquad \text{(viii)}$$

$$a_3 = -(RC\omega^2)/[1 + (RC\omega)^2]$$

Substituting $\{a_1, a_2, a_3\}$ from Eq. (viii) into Eq. (v) and combining the result with Eq. (iv) gives

$$i(t) = \mathcal{L}^{-1}\left\{\frac{V_0}{R}\left[\frac{a_1}{s + \dfrac{1}{RC}} + \frac{a_2 s + a_3}{s^2 + \omega^2}\right]\right\}$$

$$= \mathcal{L}^{-1}\left\{\frac{V_0}{R}\frac{1}{1 + (RC\omega)^2}\left[\frac{1}{s + \dfrac{1}{RC}} + \frac{(RC\omega)^2 s}{s^2 + \omega^2} - \frac{RC\omega^2}{s^2 + \omega^2}\right]\right\} \qquad \text{(ix)}$$

The inverse Laplace transform on the right-hand side of Eq. (ix) can be factored to yield

$$i(t) = \frac{V_0}{R}\frac{1}{1 + (RC\omega)^2}\mathcal{L}^{-1}\left\{\frac{1}{s + \dfrac{1}{RC}}\right\} + \mathcal{L}^{-1}\left\{\frac{(RC\omega)^2 s}{s^2 + \omega^2}\right\} - \mathcal{L}^{-1}\left\{\frac{RC\omega^2}{s^2 + \omega^2}\right\} \qquad \text{(x)}$$

Further factoring of constants gives

$$i(t) = \frac{V_0}{R}\frac{1}{1 + (RC\omega)^2}\left[\mathcal{L}^{-1}\left\{\frac{1}{s + \dfrac{1}{RC}}\right\} + (RC\omega)^2\mathcal{L}^{-1}\left\{\frac{s}{s^2 + \omega^2}\right\}\right.$$

$$\left. - RC\omega\,\mathcal{L}^{-1}\left\{\frac{\omega}{s^2 + \omega^2}\right\}\right] \qquad \text{(xi)}$$

The inverse Laplace transforms in Eq. (xi) are

$$\mathcal{L}^{-1}\left\{\frac{1}{s + \dfrac{1}{RC}}\right\} = e^{-t/RC}$$

$$\mathcal{L}^{-1}\left\{\frac{s}{s^2 + \omega^2}\right\} = \cos \omega t \qquad \text{(xii)}$$

$$\mathcal{L}^{-1}\left\{\frac{\omega}{s^2 + \omega^2}\right\} = \sin \omega t$$

Substituting Eq. (xii) into Eq. (xi) gives

$$i(t) = \frac{V_0}{R} \frac{1}{1 + (RC\omega)^2} [e^{-t/RC} + (RC\omega)^2 \cos \omega t - RC\omega \sin \omega t] \qquad \text{(xiii)}$$

Equation (xiii) is the solution obtained by Jordan and Smith [1994, p. 392].

S14 CALCULUS OF VARIATIONS

EXERCISE 14.1: Show that Eq. (14.1.4) implies that $(\partial x/\partial \alpha)|_{t_1}^{t_2} = 0$.

Solution: All of the parameterized paths of integration $x(t, \alpha) = x(t, 0) + \varepsilon(\alpha)\eta(t)$ must equal $x(t, 0)$ at the end points $\{t_1, t_2\}$. This implies that

$$\begin{aligned} x(t_1, \alpha) = x(t_1, 0) \text{ so that } \varepsilon(\alpha)\eta(t_1) = 0 \\ x(t_2, \alpha) = x(t_2, 0) \text{ so that } \varepsilon(\alpha)\eta(t_2) = 0 \end{aligned} \qquad \text{(i)}$$

Equation (i) shows that $\eta(t_1) = \eta(t_2) = 0$ for an arbitrary function $\varepsilon(\alpha)$ of the parameter α. The derivative $(\partial x/\partial \alpha)|_{t_1}^{t_2}$ is

$$\left.\frac{\partial x}{\partial \alpha}\right|_{t_1}^{t_2} = \eta(t)\left.\frac{\partial \varepsilon(\alpha)}{\partial \alpha}\right|_{t_1}^{t_2} = \eta(t_2)\frac{\partial \varepsilon(\alpha)}{\partial \alpha} - \eta(t_1)\frac{\partial \varepsilon(\alpha)}{\partial \alpha} = 0$$

because $\eta(t_1) = \eta(t_2) = 0$.

EXERCISE 14.2: What is the shortest distance between two points in a plane? *Hint*: The element of arc length in a plane is $ds = \sqrt{dx^2 + dy^2} = \sqrt{1 + (dy/dx)^2}\,dx$.

Solution: The shortest distance between two points in a plane can be determined using the calculus of variations. The element of arc length in a plane is

$$ds = \sqrt{dx^2 + dy^2} = \sqrt{1 + \left(\frac{dy}{dx}\right)^2}\,dx \qquad \text{(i)}$$

The total length of the curve is the summation of arc lengths between end points of the curve. For infinitesimal arc length, the summation is given by the integral

$$L = \int_{s_1}^{s_2} ds = \int_{x_1}^{x_2} \sqrt{1 + \left(\frac{dy}{dx}\right)^2}\,dx \qquad \text{(ii)}$$

The shortest distance between two points in a plane is determined as the extremum of Eq. (ii). The function f_1 for use in Eq. (14.1.8) is determined from

Eq. (ii) to be

$$f_1 = \sqrt{1 + \left(\frac{dy}{dx}\right)^2} = \sqrt{1 + \dot{y}^2}, \ \dot{y} = \frac{dy}{dx}$$

The condition for an extremum is determined using Eq. (14.1.8), which has the form

$$\frac{\partial f_1}{\partial y} - \frac{d}{dx}\left(\frac{\partial f_1}{\partial \dot{y}}\right) = 0$$

for the variables being used in this example. Calculating the derivatives gives

$$\frac{\partial f_1}{\partial y} = \frac{\partial \sqrt{1 + \dot{y}^2}}{\partial y} = 0$$

$$-\frac{d}{dx}\left(\frac{\partial f_1}{\partial \dot{y}}\right) = -\frac{d}{dx}\left(\frac{\partial \sqrt{1 + \dot{y}^2}}{\partial \dot{y}}\right) = -\frac{d}{dx}\left(\frac{\dot{y}}{\sqrt{1 + \dot{y}^2}}\right)$$

(iii)

Substituting Eq. (iii) into Eq. (14.1.8) gives

$$\frac{d}{dx}\left(\frac{\dot{y}}{\sqrt{1 + \dot{y}^2}}\right) = 0$$

or

$$\frac{\dot{y}}{\sqrt{1 + \dot{y}^2}} = k_1 \tag{iv}$$

where k_1 is a constant. Equation (iv) is satisfied if $\dot{y} = k_2$, where k_2 is a constant that is related to k_1 by $k_2/\sqrt{1 + k_2^2} = k_1$ from Eq. (iv). The equation $\dot{y} = k_2 = dy/dx$ can be integrated to give an equation for a straight line:

$$y = k_2 x + y_0 \tag{v}$$

where y_0 is the value of y at $x = 0$. The shortest distance between two points in a plane is a straight line.

EXERCISE 14.3: The kinetic energy and potential energy of a harmonic oscillator (HO) in one space dimension are $T_{HO} = \frac{1}{2}m\dot{q}^2$ and $V_{HO} = \frac{1}{2}mq^2$, respectively, where m is the constant mass of the oscillating object and k is the spring constant. Write the Lagrangian for the harmonic oscillator and use it in the Euler–Lagrange equation to determine the force equation for the harmonic oscillator.

Solution: The Lagrangian for the harmonic oscillator is the difference between kinetic energy and potential energy, or

$$L_{HO} = T_{HO} - V_{HO} = \frac{1}{2}m\dot{q}^2 - \frac{1}{2}mq^2 \tag{i}$$

The force on the oscillating mass is

$$F_{HO} = \frac{\partial L_{HO}}{\partial q} = -kq \tag{ii}$$

and the momentum is

$$p_{HO} = \frac{\partial L_{HO}}{\partial \dot{q}} = -m\dot{q} \tag{iii}$$

Substituting the derivatives in Eqs. (ii) and (iii) into the Euler–Lagrange equation gives

$$\frac{\partial L_{HO}}{\partial q} - \frac{d}{dt}\left(\frac{\partial L_{HO}}{\partial \dot{q}}\right) = -kq - \frac{d}{dt}(-m\dot{q}) = -kq + m\ddot{q} = 0 \tag{iv}$$

The resulting force equation for the harmonic oscillator is obtained by rearranging Eq. (iv) to yield

$$m\ddot{q} = m\frac{d^2q}{dt^2} = -kq \tag{v}$$

EXERCISE 14.4: Solve Eq. (14.2.18). *Hint*: Perform a variable substitution by introducing a parameter θ that lets you substitute $z = a\sin^2\theta = (a/2)(1 - \cos 2\theta)$ for z.

Solution: We solve Eq. (14.2.18) by introducing a parameter θ that lets us write

$$z = a\sin^2\theta = \frac{a}{2}(1 - \cos 2\theta) \tag{i}$$

Substituting Eq. (i) into Eq. (14.2.18) gives

$$\frac{dz}{dx} = \sqrt{\frac{a-z}{z}} = \cot\theta \tag{ii}$$

The integral of Eq. (ii) is

$$x = \int \frac{dz}{\cot\theta} = \int \tan\theta \, dz \tag{iii}$$

The differential dz can be written in terms of θ as

$$dz = 2a\sin\theta\cos\theta \, d\theta \tag{iv}$$

where we have taken the differential of Eq. (i). Substituting Eq. (iv) into Eq. (iii) gives

$$x = \int \tan\theta \, dz = 2a \int \tan\theta \sin\theta \cos\theta \, d\theta = \frac{a}{2}(2\theta - \sin 2\theta) + K \qquad \text{(v)}$$

where K is an integration constant. Collecting Eqs. (i) and (v) lets us write the parameterized solution of the brachistochrone problem as

$$x = \frac{a}{2}(2\theta - \sin 2\theta) + K$$
$$z = \frac{a}{2}(1 - \cos 2\theta)$$
$$\text{(vi)}$$

The equations in Eq. (vi) are the equations for a cycloid. The constants of integration $\{a, K\}$ are determined by fitting Eq. (vi) to the end points.

EXERCISE 14.5: Solve Eq. (14.3.7) for the function of two dependent variables $f_2(x_1, x_2, \dot{x}_1, \dot{x}_2, t) = (m/2)(\dot{x}_1^2 + \dot{x}_2^2) - (k/2)(x_1 + x_2)^2$ and $\{m, k\}$ are constants.

Solution: Equation (14.3.7) for two dependent variables is

$$\frac{\partial f_2}{\partial x_i} - \frac{d}{dt}\left(\frac{\partial f_2}{\partial \dot{x}_i}\right) = 0 \quad \text{for} \quad i = 1, 2$$

The derivatives are

$$\frac{\partial f_2}{\partial x_1} = -k(x_1 + x_2), \quad \frac{\partial f_2}{\partial x_2} = -k(x_1 + x_2)$$

and

$$\frac{d}{dt}\left(\frac{\partial f_2}{\partial \dot{x}_1}\right) = m\ddot{x}_1, \quad \frac{d}{dt}\left(\frac{\partial f_2}{\partial \dot{x}_2}\right) = m\ddot{x}_2$$

The two equations resulting from Eq. (14.3.7) are

$$m\ddot{x}_1 + k(x_1 + x_2) = 0, \quad m\ddot{x}_2 + k(x_1 + x_2) = 0$$

EXERCISE 14.6: Find the Euler–Lagrange equations with constraints for the function of three dependent variables $f_3(r, \theta, z) = (m/2)[\dot{r}^2 + r^2\dot{\theta}^2 + \dot{z}^2] - mgz$ subject to the constraint $r^2 = cz$, where $\{m, g, c\}$ are constants. Note that f_3 is expressed in cylindrical coordinates $\{r, \theta, z\}$.

Solution: In this case we have $n = 3$ dependent variables and $m = 1$ constraint. The constraint may be written in the form

$$g(r, \theta, z) = r^2 - cz = 0 \tag{i}$$

The derivatives corresponding to Eq. (14.4.2) are

$$\frac{\partial g}{\partial r} = 2r, \quad \frac{\partial g}{\partial \theta} = 0, \quad \frac{\partial g}{\partial z} = -c \tag{ii}$$

The Euler–Lagrange equations with one undetermined multiplier are

$$\frac{\partial f_3}{\partial r} - \frac{d}{dt}\left(\frac{\partial f_3}{\partial \dot{r}}\right) + \lambda \frac{\partial g}{\partial r} = 0$$

$$\frac{\partial f_3}{\partial \theta} - \frac{d}{dt}\left(\frac{\partial f_3}{\partial \dot{\theta}}\right) + \lambda \frac{\partial g}{\partial \theta} = 0 \tag{iii}$$

$$\frac{\partial f_3}{\partial z} - \frac{d}{dt}\left(\frac{\partial f_3}{\partial \dot{z}}\right) + \lambda \frac{\partial g}{\partial z} = 0$$

Substituting $f_3(r, \theta, z) = (m/2)[\dot{r}^2 + r^2\dot{\theta}^2 + \dot{z}^2] - mgz$ and Eq. (ii) into Eq. (iii) gives

$$m\ddot{r} - mr\dot{\theta}^2 - 2r\lambda = 0$$

$$mr^2\ddot{\theta} + 2mr\,\dot{r}\,\dot{\theta} = 0 \tag{iv}$$

$$m\ddot{z} + mg + c\lambda = 0$$

Equation (iv) is the set of equations of motion for a constant mass m sliding in a gravitational field with constant acceleration g on the inner surface of a paraboloid $r^2 = cz$ [McCuskey, 1959, p. 65].

S15 TENSOR ANALYSIS

EXERCISE 15.1: Compare the set of equations in Eq. (15.1.1) with the set of equations in Eq. (15.1.2). What are the functions $\{f^1, f^2, f^3\}$? Use these functions in Eq. (15.1.3) to calculate the components of the displacement vector in the primed (rotated) coordinate system.

Solution: Comparing Eq. (15.1.1) with Eq. (15.1.2) shows that

$$x'^1 = f^1(x^1, x^2, x^3) = x^1 \cos\theta + x^2 \sin\theta$$

$$x'^2 = f^2(x^1, x^2, x^3) = -x^1 \sin\theta + x^2 \cos\theta$$

$$x'^3 = f^3(x^1, x^2, x^3) = x^3$$

The partial derivatives $\partial x''^i/\partial x^j$ are

$$\frac{\partial x'^1}{\partial x^1} = \cos\theta, \quad \frac{\partial x'^1}{\partial x^2} = \sin\theta, \quad \frac{\partial x'^1}{\partial x^3} = 0$$

$$\frac{\partial x'^2}{\partial x^1} = -\sin\theta, \quad \frac{\partial x'^2}{\partial x^2} = \cos\theta, \quad \frac{\partial x'^2}{\partial x^3} = 0$$

$$\frac{\partial x'^3}{\partial x^1} = 0, \quad \frac{\partial x'^3}{\partial x^2} = 0, \quad \frac{\partial x'^3}{\partial x^3} = 1$$

since the rotation angle θ is a constant with respect to the unprimed coordinates. The components of the displacement vector in the primed (rotated) coordinate system are

$$dx'^1 = \cos\theta dx^1 + \sin\theta dx^2$$

$$dx'^2 = -\sin\theta dx^1 + \cos\theta dx^2$$

$$dx'^3 = dx^3$$

EXERCISE 15.2: Show that the set of equations in Eq. (15.1.9) is a special case of Eq. (15.1.10). *Hint*: Apply the limit $c \to \infty$ to the set of equations in Eq. (15.1.10).

Solution: Derive Eq. (15.1.9) from Eq. (15.1.10) by taking the limit $c \to \infty$:

$$x' = \frac{x - vt}{\sqrt{1 - (v/c)^2}} \approx x - vt \text{ because } v/c \to 0$$

$$y' = y$$

$$z' = z$$

$$t' = \frac{t - x\dfrac{v}{c^2}}{\sqrt{1 - (v/c)^2}} \approx t \text{ because } v/c \to 0 \text{ and } v/c^2 \to 0$$

EXERCISE 15.3: Write the transformation equations in Table 15.1 without the summation convention, that is, explicitly show the summations for each equation.

Solution: Table 15.1 becomes

Second-Rank Tensor	Transformation Equation
Contravariant	$A'^{ij} = \sum_{a=1}^{n} \sum_{b=1}^{n} \dfrac{\partial x'^i}{\partial x^a} \dfrac{\partial x'^j}{\partial x^b} A^{ab}$
Covariant	$A'_{ij} = \sum_{a=1}^{n} \sum_{b=1}^{n} \dfrac{\partial x^a}{\partial x'^i} \dfrac{\partial x^b}{\partial x'^j} A_{ab}$
Mixed	$A'^i_j = \sum_{a=1}^{n} \sum_{b=1}^{n} \dfrac{\partial x'^i}{\partial x^a} \dfrac{\partial x^b}{\partial x'^j} A^a_b$

EXERCISE 15.4: Use Eq. (15.2.8) to write the transformation equation for a tensor with two contravariant indices and a single covariant index.

Solution: In this case $n = 2$ and $m = 1$. The corresponding transformation equation is

$$A_{b_1}^{'a_1 a_2} = \frac{\partial x^{'a_1}}{\partial x^{c_1}} \frac{\partial x^{'a_2}}{\partial x^{c_2}} \frac{\partial x^{d_1}}{\partial x^{'b_1}} A_{d_1}^{c_1 c_2}$$

EXERCISE 15.5: Write Eq. (15.3.1) using the summation convention.

Solution: Equation (15.3.1) is written as

$$x \cdot x = \vec{x} \cdot \vec{x} = (x^1)^2 + (x^2)^2 + (x^3)^2 = x^i x_i = g_{ij} x^i x^j, x^j = x_j$$

using the summation convention.

EXERCISE 15.6: Show that the determinant of the metric tensor represented by the matrix in Eq. (15.3.2) is non zero.

Solution: The determinant of the matrix

$$\{g_{ij}\} = \begin{bmatrix} 1 & 0 & 0 \\ 0 & 1 & 0 \\ 0 & 0 & 1 \end{bmatrix}$$

is

$$|g_{ij}| = \begin{vmatrix} 1 & 0 & 0 \\ 0 & 1 & 0 \\ 0 & 0 & 1 \end{vmatrix} = 1$$

which is nonzero as expected.

EXERCISE 15.7: Show that Eq. (15.3.10) reduces to the famous result $E \approx mc^2$ when the momentum of the mass is small compared to the mass of the object.

Solution: Rearrange Eq. (15.3.10) so that total energy can be calculated from mass and momentum:

$$E^2 = m^2 c^4 + \left(p^1\right)^2 c^2 + \left(p^2\right)^2 c^2 + \left(p^3\right)^2 c^2 \tag{i}$$

We assume the momentum is small enough to be considered negligible compared to the mass of the object. In this case Eq. (i) reduces to

$$E \approx mc^2 \tag{ii}$$

If our reference frame is moving with the object, the object is at rest and Eq. (ii) becomes the equality

$$E = mc^2 = m_0 c^2 \tag{iii}$$

where m_0 is called the rest mass of the object.

EXERCISE 15.8: Calculate the contraction of the Kronecker delta function δ_j^i in n dimensions.

Solution: The contraction δ of the Kronecker delta function δ_j^i in n dimensions is obtained by summing the components of the Kronecker delta function when the contravariant and covariant indices are equal; thus $\delta = \sum_{i=1}^{n} \delta_i^i = n$. Notice that we have explicitly written the summation and are not using the summation convention.

EXERCISE 15.9: Calculate the line element for the metric tensors

$$[g_{ij}] = \begin{bmatrix} 1 & 0 \\ 0 & -1 \end{bmatrix} \quad \text{and} \quad [g_{ij}] = \begin{bmatrix} -\left(1 - \dfrac{2m}{x^1}\right)^{-1} & 0 \\ 0 & 1 - \dfrac{2m}{x^1} \end{bmatrix}$$

Assume the parameter m is a constant and x^1 is component 1 in a two-dimensional space with the displacement vector dx^i.

Solution: The line element for the metric

$$[g_{ij}] = \begin{bmatrix} 1 & 0 \\ 0 & -1 \end{bmatrix}$$

is

$$ds^2 = \sum_{i-1}^{2} \sum_{j=1}^{2} g_{ij}\, dx^i\, dx^j = dx^1 - dx^2$$

The line element for the metric

$$[g_{ij}] = \begin{bmatrix} -\left(1 - \dfrac{2m}{x^1}\right)^{-1} & 0 \\ 0 & 1 - \dfrac{2m}{x^1} \end{bmatrix}$$

is

$$ds^2 = \sum_{i=1}^{2} \sum_{j=1}^{2} g_{ij}\, dx^i dx^j = -\left(1 - \frac{2m}{x^1}\right)^{-1} dx^1 - \left(1 - \frac{2m}{x^1}\right) dx^2$$

EXERCISE 15.10: Show that $[ij, k] + [kj, i] = \partial g_{ik}/\partial x^j$, where $[\cdots]$ is the Christoffel symbol of the first kind.

Solution: From the definition of the Christoffel symbol we have

$$[ij, k] + [kj, i] = \frac{1}{2}\left\{\frac{\partial g_{ik}}{\partial x^j} + \frac{\partial g_{jk}}{\partial x^i} - \frac{\partial g_{ij}}{\partial x^k}\right\} + \frac{1}{2}\left\{\frac{\partial g_{ki}}{\partial x^j} + \frac{\partial g_{ji}}{\partial x^k} - \frac{\partial g_{kj}}{\partial x^i}\right\} \quad (i)$$

Applying the symmetry relation $g_{ab} = g_{ba}$ to the second term on the right-hand side of Eq. (i) gives

$$[ij, k] + [kj, i] = \frac{1}{2}\left\{\frac{\partial g_{ik}}{\partial x^j} + \frac{\partial g_{jk}}{\partial x^i} - \frac{\partial g_{ij}}{\partial x^k}\right\} + \frac{1}{2}\left\{\frac{\partial g_{ik}}{\partial x^j} + \frac{\partial g_{ij}}{\partial x^k} - \frac{\partial g_{jk}}{\partial x^i}\right\} \quad (ii)$$

Combining like terms in Eq. (ii) gives the desired result $[ij,k] + [kj,i] = \partial g_{ik}/\partial x^j$.

S16 PROBABILITY

EXERCISE 16.1: Calculate $P_{5,4}$, $C_{5,2}$, $C_{5,3}$, $C_{6,6}$, and $C_{6,0}$.

Solution: $P_{5,4} = 120$, $C_{5,2} = 10$, $C_{5,3} = 10$, $C_{6,6} = 1$, and $C_{6,0} = 1$.

EXERCISE 16.2: Four fair coins are each flipped once. Use a probability tree diagram to determine the probability that at least two coins will show heads.

Solution: Prepare a probability tree:

Coin #1	#2	#3	#4	
H	H	H	H	✓
			T	✓
		T	H	✓
			T	✓
	T	H	H	✓
			T	✓
		T	H	✓
			T	
T	H	H	H	✓
			T	✓
		T	H	✓

			T	
	T	H	H	✓
			T	
		T	H	
			T	

There are $2^4 = 16$ possible outcomes. The number of outcomes that have at least two heads are found by counting the outcomes that are labeled with a ✓. The probability is $11/16$.

EXERCISE 16.3: Two unbiased six-sided dice, one red and one green, are tossed and the number of dots appearing on their upper faces are observed. The results of a dice roll can be expressed as the ordered pair (g, r), where g denotes the green die and r denotes the red die.

(a) What is the sample space S?

(b) What is the probability of throwing a seven?

(c) What is the probability of throwing a seven or a ten?

(d) What is the probability that the red die shows a number less than or equal to three and the green die shows a number greater than or equal to five?

(e) What is the probability that the green die shows a one, given that the sum of the numbers on the two dice is less than four?

Solution: (a) The sample space S consists of all possible dice rolls. In this case, the sample space has 36 elements given by $S = \{(1, 1), (1, 2), \ldots, (6, 5), (6, 6)\}$.

(b) Let A denote the set of all dice rolls that add up to seven. We express A as the union of ordered pairs $\{(1, 6)\} \cup \{(2, 5)\} \cup \{(3, 4)\} \cup \{(4, 3)\} \cup \{(5, 2)\} \cup \{(6, 1)\}$. The probability of A is $P(A) = (1/36) + (1/36) + (1/36) + (1/36) + (1/36) + (1/36) = 1/6$.

(c) Let A denote the event "throwing a seven" and B denote the event "throwing a ten." Then the probability of A is $P(A) = 1/6$, and the probability of B is $P(B) = 1/12$. The probability of throwing a seven or a ten is $P(A \cup B) = P(A) + P(B) = (1/6) + (1/12) = 1/4$.

(d) Let C denote the event "red die shows number ≤ 3" and D denote the event "green die shows number ≥ 5." The elements of the set $C \cap D$ are $\{(1, 5), (2, 5), (3, 5), (1, 6), (2, 6), (3, 6)\}$. The probability $P(C \cap D)$ is the number of elements in $C \cap D$ divided by the total number of elements, thus $P(C \cap D) = 6/36$.

The independence of the events is evaluated as follows: The probability $P(C)$ of event C and the probability $P(D)$ of event D are $P(C) = 18/36$ and $P(D) = 12/36$, respectively. We have the product $P(C)P(D) = (1/2)(1/3) = 1/6 = P(C \cap D)$, which demonstrates that events C and D are independent.

(e) Let A denote the event "green die shows 1" and B denote the event "sum of numbers on dice < 4." The elements of event A are $\{(1, 1), (1, 2), (1, 3), (1, 4), (1, 5), (1, 6)\}$, and the elements of event B are $\{(1, 1), (1, 2), (2, 1)\}$. The intersection of elements is $A \cap B = \{(1, 1), (1, 2)\}$. Using the multiplicative rule, we find $P(A|B) = P(A \cap B)/P(B) = (2/36)(3/36) = 2/3$.

S17 PROBABILITY DISTRIBUTIONS

EXERCISE 17.1: What is the probability that x, y are in the intervals $1 \le x \le 2$ and $1 \le y \le 3$ for the joint probability distribution function $F(x, y) = (1 - e^{-x})(1 - e^{-y})$
for $x, y > 0$. $F(x, y)$ has the properties $F(0, 0) = 0$ and $F(\infty, \infty) = 1$, and $\rho(x, y) = e^{-x}e^{-y}$.

Solution: The probability is

$$P(1 \le x \le 2, 1 \le y \le 3) = \int_1^2 \int_1^3 \rho(x, y)dx\,dy = \int_1^2 \int_1^3 e^{-x}e^{-y}dx\,dy$$

$$= \int_1^2 e^{-x}dx \int_1^3 e^{-y}dy = (-e^{-2} + e^{-1})(-e^{-3} + e^{-1})$$

$$= e^{-5} - e^{-3} - e^{-4} + e^{-2}$$

EXERCISE 17.2: Use the conditional probability $\rho(x|y) = (x + 2y)/(\frac{1}{2} + 2y)$ defined in the intervals $0 \le x \le 1$ and $0 \le y \le 1$ to determine the probability that x will have a value less than $\frac{1}{2}$ given $y = \frac{1}{2}$.

Solution: The probability that x will have a value less than $\frac{1}{2}$ given $y = \frac{1}{2}$ is the cumulative probability

$$F(x|y) = F\left(\frac{1}{2}\bigg|\frac{1}{2}\right) = \int_0^{1/2} \left[(x + 2y) \bigg/ \left(\frac{1}{2} + 2y\right)\right] dx$$

$$= \int_0^{1/2} \left[\frac{2(x + 1)}{3}\right] dx = \frac{5}{12}$$

EXERCISE 17.3: Determine the statistical independence of the variables x and y for the joint probability density $\rho(x, y) = x^2 y^2 / 81$ by calculating the marginal probability densities. The variables x and y are defined in the intervals $0 < x < 3$ and $0 < y < 3$. Note the difference between this exercise and the intervals defined in the preceding example.

Solution: The marginal density for x is

$$\rho(x) = \int_0^3 \frac{x^2 y^2}{81} \, dy = \frac{1}{9} x^2$$

and the marginal density for y is

$$\rho(y) = \int_0^3 \frac{x^2 y^2}{81} \, dx = \frac{1}{9} y^2$$

Taking the product of marginal densities gives

$$\rho(x)\rho(y) = \left(\frac{1}{9}\right)^2 x^2 y^2 = \rho(x, y)$$

The equality of the product of marginal probability densities and the joint probability density demonstrates that x and y are independent.

EXERCISE 17.4: Calculate the first moment, or mean, of the discrete distribution

$$f(x) = \binom{N}{x} p^x (1 - p)^{N-x}$$

with the binomial coefficient $\binom{N}{x}$ for $0 \le x \le N$, and p a constant.

Solution: The distribution $f(x)$ is known as the *binomial distribution*. The first moment or mean of $f(x)$ is given by the summation

$$\bar{x} = \sum_{x=0}^N x f(x) = \sum_{x=0}^N x \binom{N}{x} p^x (1 - p)^{N-x}$$

To evaluate the mean, we first note that $f(x)$ is normalized, that is,

$$1 = \sum_{x=0}^N f(x) = \sum_{x=0}^N \binom{N}{x} p^x (1 - p)^{N-x}$$

The derivative of the normalization condition with respect to p gives

$$0 = \frac{\partial}{\partial p} \sum_{x=0}^{N} f(x) = \sum_{x=0}^{N} \binom{N}{x} [xp^{x-1}(1-p)^{N-x} - (N-x)p^x(1-p)^{N-x-1}]$$

Rearranging gives

$$\sum_{x=0}^{N} \binom{N}{x} [xp^{x-1}(1-p)^{N-x}] = \sum_{x=0}^{N} \binom{N}{x} [(N-x)p^x(1-p)^{N-x-1}]$$

$$= N \sum_{x=0}^{N} \binom{N}{x} [p^x(1-p)^{N-x-1}]$$

$$- \sum_{x=0}^{N} \binom{N}{x} [xp^x(1-p)^{N-x-1}]$$

Multiplying by $p(1-p)$ and rearranging further gives

$$\sum_{x=0}^{N} x \binom{N}{x} [(1-p)p^x(1-p)^{N-x} + pp^x(1-p)^{N-x}] = Np \sum_{x=0}^{N} \binom{N}{x} [p^x(1-p)^{N-x}]$$

The left-hand side can be simplified by canceling terms in the square bracket to obtain

$$\sum_{x=0}^{N} x \binom{N}{x} [p^x(1-p)^{N-x}] = \bar{x} = Np \sum_{x=0}^{N} \binom{N}{x} [p^x(1-p)^{N-x}]$$

where we have used the definition of the mean. If we now apply the normalization condition to the right-hand side, we find the simple result $\bar{x} = Np$ as the mean of the binomial distribution.

EXERCISE 17.5: Suppose we are given the joint probability density $p(x, y, z) = (x+y)e^{-z}$ for three random variables defined in the intervals $0 < x < 1$, $0 < y < 1$, and $0 < z$.

(a) Find the joint probability density $p(x, y)$.

(b) Find the marginal probability density $p(y)$.

Solution: (a) The joint probability density $\rho(x, y)$ is found by integrating over the variable z; thus

$$\rho(x, y) = \int_0^\infty \rho(x, y, z)\, dz = \int_0^\infty (x + y)\, e^{-z} dz = (x + y) \int_0^\infty e^{-z} dz = x + y$$

(b) The marginal probability density $\rho(y)$ is found by integrating the joint probability density $\rho(x, y)$ over the variable x; thus

$$\rho(y) = \int_0^1 \rho(x + y)\, dx = \int_0^1 (x + y)\, dx = y + \frac{1}{2}$$

EXERCISE 17.6: Determine the expectation value $E(x)$ using the probability density for the normal distribution.

Solution: The expectation value is

$$E(x) = \int_{-\infty}^\infty x\rho(x)\, dx = \frac{1}{\sqrt{2\pi}\sigma} \int_{-\infty}^\infty x \exp\left[-\frac{(x - \mu)^2}{2\sigma^2}\right] dx$$

The integral is solved as follows. Define the new variable

$$z = \frac{x - \mu}{\sqrt{2}\sigma},\ dz = \frac{dx}{\sqrt{2}\sigma}$$

so that the integral becomes

$$E(x) = \frac{1}{\sqrt{2\pi}\,\sigma} \int_{-\infty}^\infty \left(\sqrt{2}\,\sigma z + \mu\right) \exp\left[-z^2\right] \sqrt{2}\,\sigma dz$$

$$= \frac{\sqrt{2}\,\sigma}{\sqrt{\pi}} \int_{-\infty}^\infty z \exp\left[-z^2\right] dz + \frac{\mu}{\sqrt{\pi}} \int_{-\infty}^\infty \exp\left[-z^2\right] dz$$

The first integral on the right-hand side of the equation is an integral of an odd function evaluated over a symmetric interval and is therefore zero. The result is

$$E(x) = \frac{\mu}{\sqrt{\pi}} \int_{-\infty}^\infty \exp\left[-z^2\right] dz = \mu$$

since

$$\int_{-\infty}^\infty \exp\left[-z^2\right] dz = \sqrt{\pi}$$

S18 STATISTICS

EXERCISE 18.1: Prepare a histogram of the grouped data in Table 18.1.

Solution: The histogram is shown in the following figure.

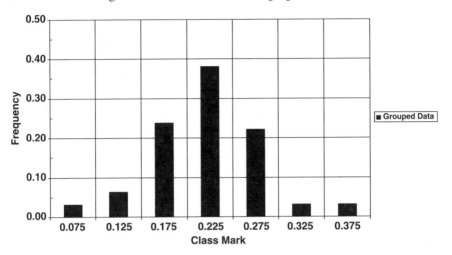

EXERCISE 18.2: What is the mean of the grouped data in Table 18.1?

Solution: The mean for grouped data is

$$\bar{x} = \frac{1}{n}\sum_{i=1}^{k} f_i x_i$$

$$= \frac{2(0.075) + 4(0.125) + 15(0.175) + 24(0.225) + 14(0.275) + 2(0.325) + 2(0.375)}{63}$$

$$= 0.221$$

EXERCISE 18.3: Show that the power function $y = ax^b$ can be recast in linear form.

Solution: The power function $y = ax^b$ is recast in linear form by taking the logarithm

$$\log y = \log a + b \log x$$

For a power function fit by the method of least squares, the values $\log a$ and b are obtained by fitting a straight line to the set or ordered pairs $\{(\log x_i,\ \log y_i)\}$. The slope of the straight line on a log–log plot is b and the corresponding intercept of the straight line is $\log a$.

REFERENCES

Abramowitz, M.A. and I.A. Stegun (1972): *Handbook of Mathematical Functions*, 9th printing, Dover, New York.

Arfken, G.B. and H.J. Weber (2001): *Mathematical Methods for Physicists*, 5th ed., Elsevier-Academic Press, New York.

Aziz, K. and A. Settari (1979): *Petroleum Reservoir Simulation*, Applied Science Publishers, London.

Beltrami, E. (1987): *Mathematics for Dynamic Modeling*, Academic Press, Boston.

Berberian, S.K. (1992): *Linear Algebra*, Oxford University Press, New York.

Blaisdell, E.A. (1998): *Statistics in Practice*, 2nd ed., Saunders College Publishing, Fort Worth, TX.

Boas, M. (2005): *Mathematical Methods in the Physical Sciences*, 3rd ed., Wiley, Hoboken, NJ.

Breckenbach, E.F., I. Drooyan, and M.D. Grady (1985): *College Algebra*, 6th ed., Wadsworth, Belmont, CA.

Burger, J.M. (1948): A mathematical model illustrating the theory of turbulence, *Adv. Appl. Mechanics* **1**: 171–199.

Carnahan, B., H.A. Luther, and J.O. Wilkes (1969): *Applied Numerical Methods*, Wiley, Hoboken, NJ.

Chapra, S.C. and R.P. Carnale (2002): *Numerical Methods for Engineers*, McGraw-Hill, Boston.

Chow, T.L. (2000): *Mathematical Methods for Physicists, A Concise Introduction*, Cambridge University Press, Cambridge, UK.

Collins, R.E. (1977): The mathematical basis of quantum mechanics, *Lett. Nuovo Cim.* **18**: 581–584.

Collins, R.E. (1979): The mathematical basis of quantum mechanics: II, *Lett. Nuovo Cim.* **25**: 473–475.

Collins, R.E. (1985): Characterization of heterogeneous reservoirs, Research & Engineering Consultants, and private communication.

Collins, R.E. (1999): *Mathematical Methods for Physicists and Engineers*, 2nd corrected ed., Dover, Mineola, NY.

Corduneanu, C. (1991): *Integral Equations and Applications*, Cambridge University Press, Cambridge, UK.

Devaney, R.L. (1992): *A First Course in Chaotic Dynamical Systems*, Addison-Wesley, Reading, MA.

DeVries, P.L. (1994): *A First Course in Computational Physics*, Wiley, Hoboken, NJ.

Eddington, A.S. (1960): *The Mathematical Theory of Relativity*, Cambridge University Press, Cambridge, UK.

Edwards, C.H. Jr. and D.E. Penney (1989): *Elementary Differential Equations with Boundary Value Problems*, 2nd ed., Prentice-Hall, Englewood Cliffs, NJ.

Fanchi, J.R. (1983): Multidimensional numerical dispersion, *Soc. Pet. Eng. J.* 143–151.

Fanchi, J.R. (1993): *Parametrized Relativistic Quantum Theory*, Kluwer, Dordrecht.

Fanchi, J.R. (1997): *Principles of Applied Reservoir Simulation*, Gulf Publishers, Houston, TX.

Fröhner, F.H. (1998): Missing link between probability theory and quantum mechanics: the Riesz–Fejér theorem, *Z. Naturforsch.* **53a**: 637–654.

Gluckenheimer, J. and P. Holmes (1986): *Nonlinear Oscillations, Dynamical Systems, and Bifurcations of Vector Fields*, Springer, New York.

Goldstein, H., C. Poole, and J. Safko (2002): *Classical Mechanics*, 3rd ed., Addison-Wesley, New York.

Gustafson, R.D. and P.D. Frisk (1980): *College Algebra*, Brooks/Cole, Monterey, CA.

Hirsch, M.W. and S. Smale (1974): *Differential Equations, Dynamical Systems, and Linear Algebra*, Academic Press, New York.

Jackson, J.D. (1962): *Classical Electrodynamics*, Wiley, Hoboken, NJ.

Jensen, R.V. (1987): Classical chaos, *Am. Sci.* **75**: 168–181.

John, F. (1978): *Partial Differential Equations*, 3rd ed., Springer-Verlag, New York.

Jordan, D.W. and P. Smith (1994): *Mathematical Techniques: An Introduction for the Engineering, Physical and Mathematical Sciences*, Oxford University Press, Oxford, UK.

Kreider, D.L., R.G. Kuller, and D.R. Ostberg (1968): *Elementary Differential Equations*, Addison-Wesley, Reading, MA.

Kreyszig, E. (1999): *Advanced Engineering Mathematics*, 8th ed., Wiley, Hoboken, NJ.

Kurtz, M. (1991): *Handbook of Applied Mathematics for Engineers and Scientists*, McGraw-Hill, New York.

Lapidus, L. and G.F. Pinder (1982): *Numerical Solution of Partial Differential Equations in Science and Engineering*, Wiley, Hoboken, NJ.

Larsen, R.J. and M.L. Marx (1985): *An Introduction to Probability and Its Applications*, Prentice-Hall, Englewood Cliffs, NJ.

Larson, R.E., R.P. Hostetler, and B.H. Edwards (1990): *Calculus with Analytic Geometry*, 4th ed., D.C. Heath, Lexington, MA.

Lebedev, L.P. and M.J. Cloud (2003): *Tensor Analysis*, World Scientific, Singapore.

Lichtenberg, A.J. and M.A. Lieberman (1983): *Regular and Stochastic Motion*, Springer-Verlag, New York.

Lipschutz, S. (1968): *Linear Algebra*, McGraw-Hill, New York.

Lomen, D. and J. Mark (1988): *Differential Equations*, Prentice-Hall, Englewood Cliffs, NJ.

Mandelbrot, B.B. (1967): How long is the coast of Britain? Statistical self-similarity and fractional dimension, *Science* **155**: 636–638.

Mandelbrot, B.B. (1983): *The Fractal Geometry of Nature*, Freeman, New York.

May, R.M. (1976): Simple mathematical models with very complicated dynamics, *Nature* **261**: 459ff.

McCuskey, S.W. (1959): *An Introduction to Advanced Dynamics*, Addison-Wesley, Reading, MA.

Mendenhall, W. and R.J. Beaver (1991): *Introduction to Probability and Statistics*, PWS-Kent Publishers, Boston, MA.

Miller, I. and M. Miller (1999): *John E. Freund's Mathematical Statistics*, 6th ed., Prentice-Hall, Upper Saddle River, NJ.

Nicholson, W.K. (1986): *Elementary Linear Algebra with Applications*, PWS-Kent Publishers, Boston, MA.

Parker, T.S. and L.O. Chua (1987): Chaos: a tutorial for engineers, *Proc. IEEE* **75**: 982–1008.

Peaceman, D.W. (1977): *Fundamentals of Numerical Reservoir Simulation*, Elsevier, New York.

Press, W.H., S.A. Teukolsky, W.T. Vetterling, and B.P. Flannery (1992): *Numerical Recipes*, 2nd ed., Cambridge University Press, New York.

Rich, B. (1973): *Modern Elementary Algebra*, McGraw-Hill, New York.

Riddle, D.F. (1992): *Analytic Geometry*, 5th ed., Wadsworth, Belmont, CA.

Rudin, W. (1991): *Functional Analysis*, 2nd ed., McGraw-Hill, New York.

Spiegel, M.R. (1963): *Advanced Calculus*, McGraw-Hill, New York.

Stewart, J., L. Redlin, and S. Watson (1992): *College Algebra*, Brooks/Cole, Pacific Grove, CA.

Stewart, J. (1999): *Calculus*, 4th ed., Brooks/Cole, Pacific Grove, CA.

Swokowski, E.W. (1986): *Algebra and Trigonometry with Analytic Geometry*, Prindle, Weber & Schmidt, Boston, MA.

Tatham, R.H. and M.D. McCormack (1991): *Multicomponent Seismology in Petroleum Exploration*, Society of Exploration Geophysicists, Tulsa, OK.

Telford, W.M., L.P. Geldart, and R.E. Sheriff (1990): *Applied Geophysics*, 2nd ed., Cambridge University Press, Cambridge, UK.

Thomas, G.B. (1972): *Calculus and Analytic Geometry*, alternate ed., Addison-Wesley, Reading, MA.

Walpole, R.E. and R.H. Myers (1985): *Probability and Statistics for Engineers and Scientists*, 3rd ed., Macmillan Publishing, New York.

Yilmaz, Ö. (2001): *Seismic Data Analysis*, Volume I, Society of Exploration Geophysicists, Tulsa, OK.

Zwillinger, D. (2003): *CRC Standard Mathematical Tables and Formulae*, 31st ed., Chapman & Hall/CRC, Boca Raton, FL.

INDEX

Math Refresher for Scientists and Engineers, Third Edition By John R. Fanchi
Copyright © 2006 John Wiley & Sons, Inc.